Rapid Manufacturing

Springer
London
Berlin
Heidelberg
New York
Barcelona
Hong Kong
Milan
Paris
Singapore
Tokyo

D.T. Pham and S.S. Dimov

Rapid Manufacturing

The Technologies and Applications of Rapid Prototyping and Rapid Tooling

With 201 Figures

Springer

D.T. Pham, BE, PhD, DEng, FREng, CEng, FIEE
S.S. Dimov, Dipl Eng, PhD
Manufacturing Engineering Centre, School of Engineering, Cardiff University, Cardiff, CF24 0YF

ISBN 1-85233-360-X Springer-Verlag London Berlin Heidelberg

British Library Cataloguing in Publication Data
Pham, D.T. (Duc Truong), 1952-
Rapid manufacturing : the technologies and applications of rapid prototyping and rapid tooling
1.Design, Industrial 2.Prototypes, Engineering
3.Manufacturing processes
I.Title II.Dimov, S. S.
658.5'7

ISBN 185233360X

Library of Congress Cataloging-in-Publication Data
Pham, D.T.
Rapid manufacturing : the technologies and applications of rapid prototyping and rapid tooling / D.T. Pham and S.S. Dimov.
p. cm.
Includes bibliographical references and index.
ISBN 1-85233-360-X (alk. paper)
1. Design, Industrial—Data processing. 2. Prototypes, Engineering. 3. CAD/CAM systems. 4. Manufacturing processes. I. Dimov, S.S., 1959- II. Title.
TS171.4 .P453 2000
658.5'7—dc21 00-063767

Typesetting: Electronic text files prepared by authors
Printed and bound by The Cromwell Press, Trowbridge, Wiltshire, UK
69/3830-543210 Printed on acid-free paper SPIN 10571891

Preface

Rapid manufacturing is a term that embraces rapid prototyping (RP) and rapid tooling (RT).

RP is an exciting new technology for quickly creating physical models and functional prototypes directly from CAD models. RT generally concerns the production of tooling using inserts. RP and RT are means for compressing the time-to-market of products and, as such, are competitiveness enhancing technologies.

The first five chapters of this book describe the characteristics, capabilities and applications of the main known RP processes. Chapter 1 introduces the history of RP and discusses CAD modelling as a prelude to physical prototyping. Chapter 2 provides an overview of different RP techniques including those still under development. Chapter 3 examines commercially available RP machines, such as systems for Stereolithography (SLA), Selective Laser Sintering (SLS), Fused Deposition Modelling (FDM), Solid Ground Curing (SGC), Laser Engineering Net Shaping (LENS) and Laminated Object Manufacturing (LOM). Chapter 4 focuses on a special class of commercial machines, the so-called "concept modellers" mainly to be found in design offices. Chapter 5 covers the varied uses of RP as a means for building functional prototypes, patterns for castings, medical models, artworks and test pieces for engineering analysis.

The next three chapters discuss RT and its applications. Indirect methods of producing soft tooling, firm tooling (or bridge tooling) and hard tooling based on RP are described in Chapter 6, while direct methods are dealt with in Chapter 7. Chapter 8 examines the technological capabilities of commercial RT processes from two major suppliers, DTM and EOS. The chapter also presents industrial case studies illustrating the use of RT in two key areas, plastics injection moulding and aluminium die-casting.

The final chapter in the book, Chapter 9, is concerned with process optimisation. It reviews the main factors affecting the accuracy of RP parts and the process-specific constraints to be considered when choosing part build orientations. The chapter describes the selection of process parameters and part orientations to achieve the best compromise between a good accuracy and surface finish and a low build cost and fast turn around.

To illustrate the principles of the machines and processes discussed, the book uses an abundance of diagrams and photographs. Some of the illustrations are reproduced on colour plates for improved clarity. These illustrations are marked with an asterisk (*) in the text.

The book was written so that no specialist technical background would be required of readers who were assumed to be engineers in design, research, development and manufacturing. Accordingly, the book places a strong emphasis on practical engineering issues, containing material derived from industrial case studies undertaken at the authors' Centre. Many of those studies formed part of collaborative research projects financed by the European Commission, the European Regional Development Fund, the Welsh Assembly and the Welsh Development Agency. The authors gratefully acknowledge the support of these organisations and that of the collaborating companies and their personnel. In particular, the authors wish to thank Mr S Smith and Dr C Bryant of the WDA, Dr D Williams formerly of Alloycast, Mr F D Marsh of GX Design Engineers Ltd and Mr I Stead of Iota Sigma for their help with many of the projects at the Centre.

Several colleagues at the Centre and the Cardiff School of Engineering have also contributed to this book. The authors are most appreciative of the permission to use material from the research of Mr C Ji, Dr R S Gault and Dr F Lacan. Dr K Dotchev, Mr A Ivanov and Dr X Wang worked on the industrial case studies reported in the book and are thanked for this, so are Mr B G Watkins and Mr V O'Hagan who assisted with the manuscript checking and Mr A Rowlands who proofread the complete text. The authors are also grateful to Dr B J Peat for his invaluable help with the final editing of the book to ensure consistency and compliance with Springer Verlag's house style.

The book includes copyrighted text and photographs. The authors are thankful to publishers Elsevier, the IEE, the IMechE and Springer Verlag, and RP machine manufacturers/suppliers/users 3D Systems, American Precision Products, Anatomics, DTM, CALM consortium, EOS GmbH, Helisys, Kira Corp, MCP, Materialise, Objet Geometries, Optomec Design, Sanders Prototype Inc, Schroff Development Corp, Soligen, Stratasys, Tritech, and Z-Corp for their permission to use this material.

Finally, the authors wish to thank Mr N Pinfield and Ms H Ransley of Springer Verlag London and Mrs A Jackson, formerly of the same company, for their patient support throughout the writing of this book.

D T Pham and S S Dimov
Manufacturing Engineering Centre
School of Engineering
Cardiff University

Contents

Chapter 1 Introduction

Global competition, customer-driven product customisation, accelerated product obsolescence and continued demands for cost savings are forcing companies to look for new technologies to improve their business processes and speed up the product development cycle. Rapid Prototyping (RP) has emerged as a key enabling technology with its ability to shorten product design and development time. RP technologies can be virtual and physical.

Virtual Prototyping (VP) is a means of carrying out the analysis and simulation of products employing digital mock-ups (3D product representations). This allows product performance to be investigated before any physical parts are built. VP is usually tightly integrated with CAD/CAM and sometimes referred to as Computer-Aided Engineering (CAE).

Physical RP builds tangible objects from computer data without the need of jigs or fixtures or NC programming. This technology has also been referred to as layer manufacturing, solid free-form fabrication, material addition manufacturing and three-dimensional printing.

This book focuses on physical RP processes and their applications. The current introduction chapter starts with an historical perspective of this technology and discusses the role of RP in Time-Compression Engineering including the available data formats and interfaces to 3D CAD modelling systems. Finally, it outlines the main stages in generating the necessary data for guiding RP processes.

1.1 Historical Perspectives

This section is based on [Beaman, 1997].

The roots of RP can be traced to two technical areas [Beaman, 1997]: topography and photosculpture.

1. *Topography*. A layered method was proposed by Blanther as early as 1890 [Blanther, 1892] for making moulds for topographical relief maps. Both positive and negative 3D surfaces were to be assembled from a series of wax plates cut along the topographical contour lines (Figure 1.1). This method was further refined by Perera [Perera, 1940], Zang [Zang, 1964] and Gaskin [Gaskin, 1973]. Matsubara [Matsubara, 1972] described a layer manufacturing process to form casting moulds. The layers of the moulds are produced from refractory particles coated with a photopolymer resin. The resin is selectively cured using light. Similarly, DiMatteo [DiMatteo, 1976] proposed a process for layer manufacturing 3D objects from contoured metallic sheets that are formed using a milling cutter. Nakagawa reported the use of lamination techniques for fabrication of blanking tools [Nakagawa et al., 1979], press forming tools [Kunieda and Nakagawa, 1984] and injection moulding tools [Nakagawa et al., 1985].

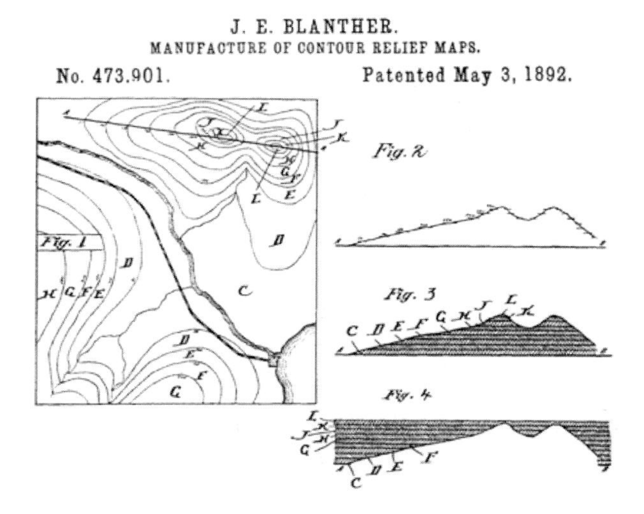

Figure 1.1 A method for making moulds for topographical relief maps [Blanther, 1892]

2. *Photosculpture.* This is a technique [Bogart, 1979] proposed in the 19th century for creating replicas of 3D objects. The technique involves photographing the object simultaneously with 24 cameras equally spaced around a circular room and then using the silhouette of each photograph to carve 1/24th of a cylindrical portion of the object. Attempts were made by other developers [Baese, 1904; Monteah, 1924] to improve the technique by alleviating the manual carving steps. Morioka [Morioka, 1935; Morioka, 1944] proposed the use of structured lighting to create contour lines of an object photographically and then using these lines to cut and build the object from sheets. In 1956, Munz [Munz, 1956] patented a layer manufacturing system for fabricating the cross-sections of a scanned object by selectively exposing a transparent photo emulsion (Figure 1.2). The system produces the layers by lowering a piston in a cylinder and adding appropriate amounts of photo emulsion and fixing agent.

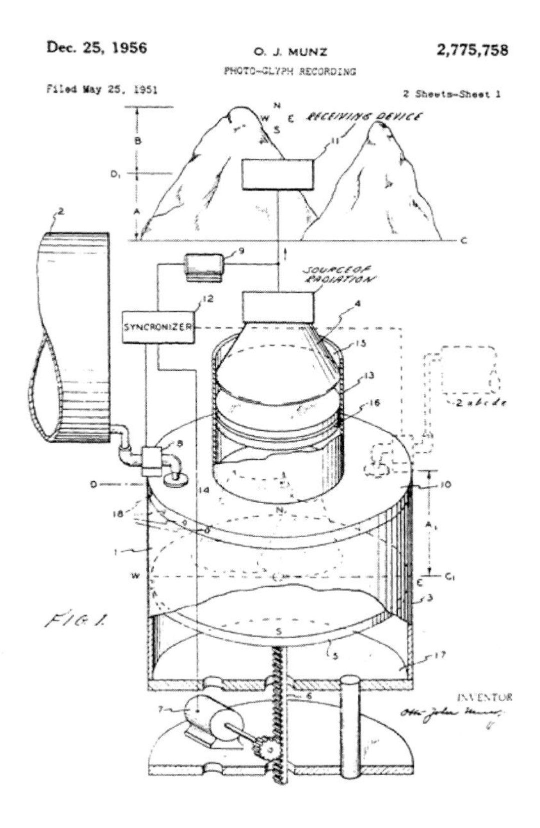

Figure 1.2 The layer manufacturing system proposed by Munz [1956]

Development work in the area of RP continued in the 1960s and 1970s and a number of patents have been filed on different methods and systems [Beaman, 1997]. These include:

- A method for fabricating objects from powdered materials by heating particles locally and fusing them together employing a laser, electron beam, or plasma beam [Ciraud, 1972].

- A process for producing plastic patterns by selective 3D polymerisation of a photosensitive polymer at the intersection of two laser beams [Swainson, 1977].

- A photopolymer RP system for building objects in layers [Kodama, 1981]. A mask is used to control the exposure of the UV source when producing a cross-section of the model.

- A system that directs a UV laser beam to a polymer layer by means of a mirror system on an x-y plotter [Herbert, 1982].

Further to this list there are numerous patents covering existing commercial RP processes. The most prominent patents as listed by Beaman [Beaman, 1997] are shown in Figure 1.3.

The significant increase in the number of commercially available RP systems of the 1990s can be explained by advances in 3D CAD Modelling, Computer-Aided Manufacturing and Computer Numerical Control. These technologies were used initially in the fast growing, highly competitive, high technology, automotive and aerospace industries, which generated added momentum. At the beginning and in the middle of the 1990s, the annual growth in sales of RP systems was approaching 40-50%. In the last few years, the same rapid growth has not continued but developments in this area still attract significant interest and in the last two years 208 new patents were filed. In 1999, sales growth was 22% and it was estimated that 3.4 million parts were built world-wide using RP technologies [Wohlers, 2000]. Another important aspect is that the application of RP has spread to other sectors of the economy (Figure 1.4). This strong and consistent growth in sales and the widespread use of the technology present very optimistic prospects for the RP industry and its future.

Name	Title	Filed	Country
Housholder	Moulding process	Dec-79	U.S.
Murutani	Optical mould method	May-84	Japan
Masters	Computer automated manufacturing process and system	Jul-84	U.S.
Andre et al.	Apparatus for making a model of an industrial part	Jul-84	France
Hull	Apparatus for making three-dimensional objects by stereolithography	Aug-84	U.S.
Pomerantz et al.	Three-dimensional mapping and modelling apparatus	Jun-86	Israel
Feygin	Apparatus and method for forming an integral object from laminations	Jun-86	U.S.
Deckard	Method and apparatus for producing parts by selective sintering	Oct-86	U.S.
Fudim	Method and apparatus for producing three-dimensional objects by photosolidification; radiating an uncured photopolymer	Feb-87	U.S.
Arcella et al.	Casting shapes	Mar-87	U.S.
Crump	Apparatus and method for creating three-dimensional objects	Oct-89	U.S.
Helinski	Method and means for constructing three-dimensional articles by particle deposition	Nov-89	U.S.
Marcus	Gas phase selective beam deposition: three-dimensional, computer-controlled	Dec-89	U.S.
Sachs et al.	Three-dimensional printing	Dec-89	U.S.
Levent et al.	Method and apparatus for fabricating three-dimensional articles by thermal spray deposition	Dec-90	U.S.
Penn	System, method, and process for making three-dimensional objects	Jun-92	U.S.

Figure 1.3 Active RP patents [Beaman, 1997]

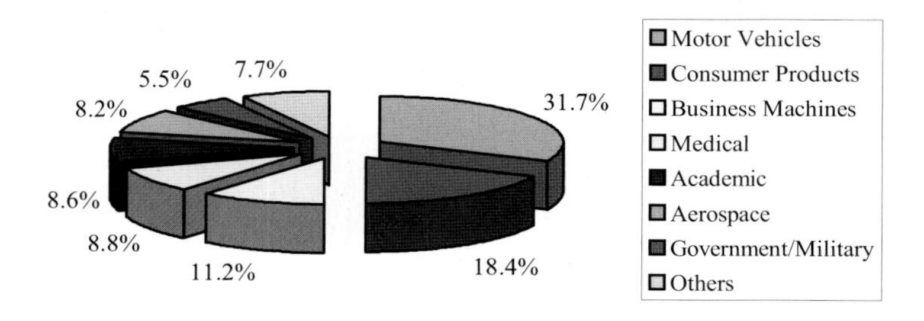

Figure 1.4 The use of RP systems in different sectors [Wohlers, 2000]

1.2 Rapid Prototyping – An Integral Part of Time Compression Engineering

Time-Compression Engineering (TCE) is also known as Concurrent Engineering. The main enabling technology behind TCE is 3D CAD modelling. Concurrency in performing different design and manufacturing activities presents an opportunity to compress the overall product development time whilst opening up possibilities to be creative by providing more time for design iterations (Figure 1.5).

Concurrent Engineering environments have evolved considerably during the last few years to integrate 3D modelling with CAM, CAE, Rapid Prototyping and Manufacturing and a number of other applications. The 3D model becomes a central component of the whole product or project information base which means that in all design, analysis and manufacturing activities the same data is utilised. There is no duplication and no misunderstanding. Product information captured in this way can be copied and re-used; it is readily available for different down-stream applications.

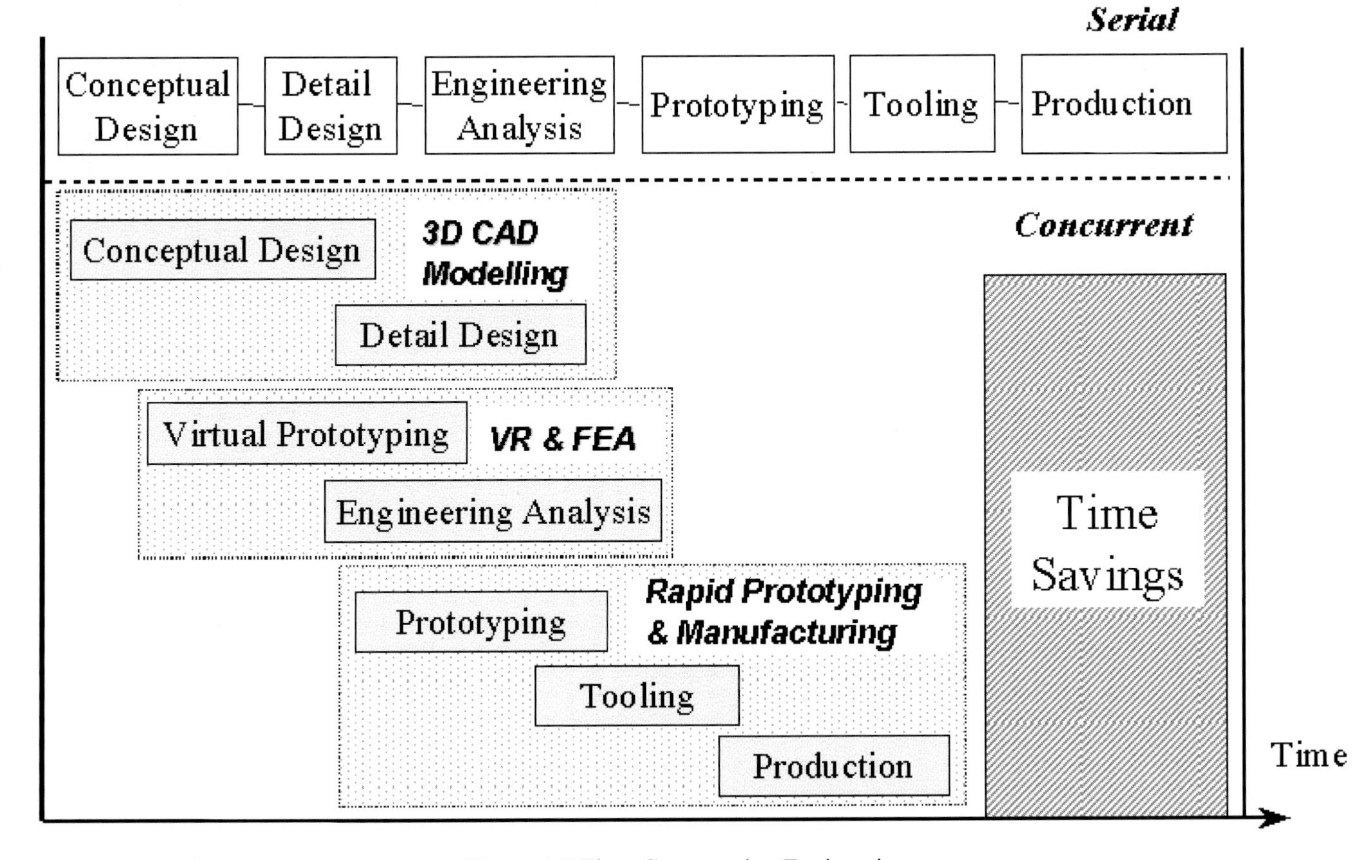

Figure 1.5 Time-Compression Engineering

3D models and virtual prototypes allow most problems with fitting to become obvious early in the product development process. Assemblies can be verified for interference as VP can be exercised through a range of tasks. Structural and thermal analysis can be performed on the same models employing CAE applications as well as simulating down-stream manufacturing processes. Ultimately, these very accurate and data rich models can be taken directly to RP and CAM applications speeding up process planning and in some cases eliminating the need for drawings.

As a gateway to RP, 3D modelling deserves special attention. A thorough understanding of different geometrical representation techniques and formats for data exchange is crucial for the successful utilisation of RP technology. The next two paragraphs review briefly the existing 3D modelling techniques and data formats used for RP purposes.

1.2.1 Geometrical Modelling Techniques

The introduction of computer-controlled fabrication systems, especially that of Numerical Controlled machine tools some 40 years ago, created the need for electronic representation of product data. The first generation of computer-aided design tools were 2D drafting systems that electronically mirror the processes traditionally conducted by draftsmen in the drawing office. 2D geometrical models consist of graphical primitives such as lines, arcs, text, symbols, and other notations required to represent engineering drawings in an electronic format.

The modelling capabilities of this first generation of CAD systems were very limited. Only basic design applications were available and these applications were far from being capable of handling real industrial design problems. The main ideas implemented in these systems were further developed and utilised in the next generation of CAD systems which had 3D modelling capabilities. Increases in product complexity and the need to integrate and/or automate various elements of design and manufacturing drove the development of these 3D CAD systems.

For geometrical models to become acceptable in a wide range of engineering applications, they should be three-dimensional, unique and complete representations of products. Such models allow the same data to be used in different engineering tasks from documentation (drafting), to engineering analysis, rapid prototyping and manufacturing. There are three basic modelling techniques available to create 3D designs: wireframe, surface and solid modelling.

1.2.1.1 Wireframe Modelling

Analogous to 2D geometrical models, *wireframe models* consist of graphical primitives defined in three-dimensional space. These models represent 3D design objects only with edges and vertices. Edges can be lines or curves. The construction of valid 3D models employing wireframe techniques is considered to be a lengthy and difficult process because of the amount of input data and command sequences needed to create them. However, wireframe models can be easily stored in engineering databases as only a small amount of computer memory is required and the data can be retrieved, edited or updated quickly. The main purpose of wireframe models is to support the creation of engineering documentation and also in some cases to serve as input data for finite element analysis. Using these models, various projections of the 3D object can be created by applying geometrical transformations to the graphical primitives.

Unfortunately, 3D designs represented by wireframe models cannot be interpreted uniquely and models of complex objects can become impossible to interpret (Figure 1.6). To overcome possible confusion, edges can be hidden, dashed, or blanked. Nevertheless, difficulties with the interpretation of the models led to the position where most 3D wireframe systems installed in companies are used in two-dimensional mode only. Wireframe models do not contain surface and volume data, which makes them unsuitable for RP purposes. In general, wireframe modelling techniques are considered natural extensions of traditional drafting methods.

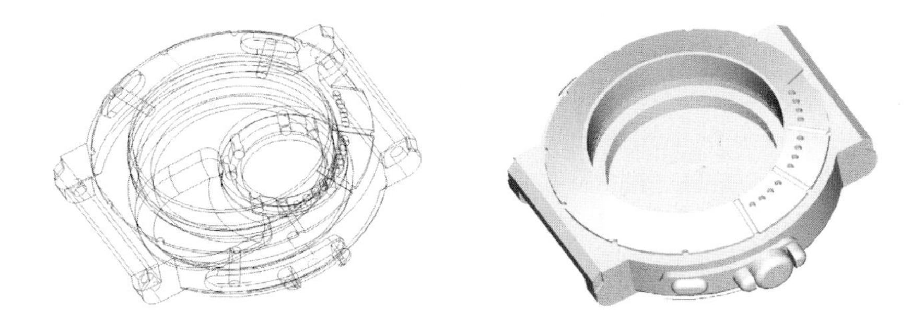

Figure 1.6 Wireframe and solid models of a watch

1.2.1.2 Surface Modelling

Wireframe models form the basis for the creation of surface models. Most existing surface modellers require wireframe primitives to generate surfaces. The user enters the vertices and edges as in wireframe models to define the boundaries of each surface. Then, using the entered data, surfaces are fitted to these edges. Different mathematical techniques can be employed to represent surfaces. For example, free-form surfaces can be represented using one of the following geometrical representation techniques: Coons, Bezier, Non Uniform Rational B-splines (NURBS), quadratic, cylindrical or spherical surfaces [Zeid, 1991].

Surface models are more complete and less ambiguous representations than their wireframe counterparts. Their geometrical databases are richer and provide information on surfaces connecting model edges. This data is sufficient for generating cutter paths for NC machining and therefore most CAM systems are based on this representation technique.

Unfortunately, surface models define only the geometry of objects, do not store any information about their topology and can only be regarded as a set of surfaces belonging to one object. Thus, if one edge is common to two surfaces this information is not stored in the model. This leads to the existence of gaps between the surfaces which means that surface models cannot define closed volumes. To use surface models for RP purposes, these gaps must be removed, which can be very difficult or even impossible.

1.2.1.3 Solid Modelling

The definition of models in solid modelling is easier than with the other two modelling techniques. Minimal input data is required and command sequences are much simpler. Most Solid Modelling packages support a Constructive Solid Geometry (CSG) user input. This user interface allows complex objects to be built from a set of predefined 3D primitives. These primitives can be either simple basic shapes such as planes, cylinders, cones, spheres, etc. or more complex solid objects created by sweeping 2D sections of wireframe entities. To define a solid model, such primitives are combined using the boolean operations of union, intersection and difference.

Solid models provide a complete and unambiguous representation of objects [Zeid, 1991]. The completeness and unambiguity of these models are due to the information stored in their databases. After a part is constructed, the solid modeller converts the input into a data structure which maintains the geometry and topology of the object. In contrast to both wireframe and surface models that store only geometrical data, solid modelling databases are complete and the models are very easy to verify.

Various representation schemes exist for storing solid modelling data but the two most popular schemes are CSG and Boundary Representation (B-Rep) [Zeid, 1991]. CSG stores objects as a tree, where the leaves are solid primitives and the interior nodes are boolean operations (Figure 1.7). The B-Rep scheme is based on the topological notion that each 3D object is bounded by a set of faces. The data is stored as sets of faces, edges and vertices that are linked together to ensure topological consistency of the model (Figure 1.8). In most solid modelling packages more than one representation scheme is supported. One is considered the primary scheme and the other representations are derived from it.

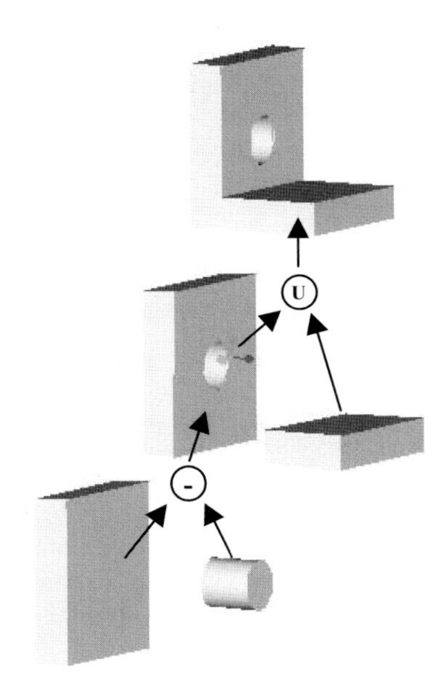

Figure 1.7 A construction tree for a solid model

Solid modelling has been acknowledged as a key element in the integration of design and manufacturing [Zeid, 1991]. Its widespread use was made possible by the large increase in computing power to cost ratio over the last 10 years. Solid modelling is now considered the most reliable way of creating 3D models for RP purposes.

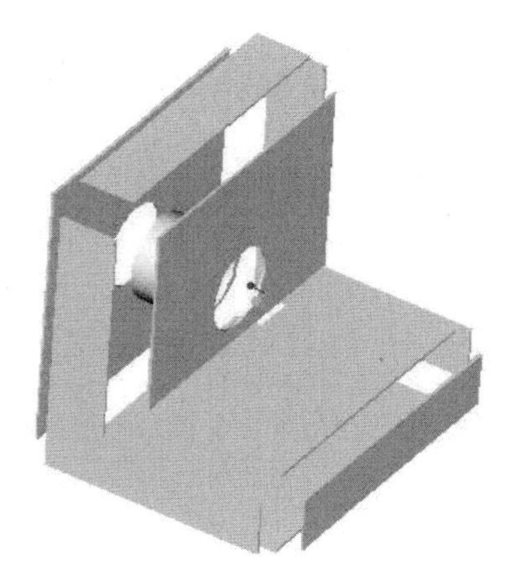

Figure 1.8 Boundary representation of a solid model

1.2.2 RP Data Formats

3D modelling packages store geometrical data employing different representation techniques and data formats. Incompatibility of these formats makes data exchange between these modellers and downstream CAM applications difficult. Two methods exist to overcome this problem, one direct and the other indirect [Zeid, 1991]. The direct solution requires the development of an interface between any two systems, which in most cases is impractical. The second solution employs neutral database structures to exchange information between the systems. Usually, these data structures are general and independent of any particular vendor. They contain the minimum required definitions to execute a particular group of applications.

In RP, the indirect approach is adopted. Several neutral formats have been proposed: STL (STereoLithography) [Jacob, 1996], SLC [Jacob, 1996], CLI (Common Layer Interface) [Brite, 1994], RPI (Rapid Prototyping Interface) [Rock and Wozny, 1991], LEAF (Layer Exchange Ascii Format) [Dolenc and Malela, 1992] and LMI (Layer Manufacturing Interface) [Kai et al., 1997]. The most commonly used format is STL which is considered de facto standard for interfacing CAD and RP systems. The other formats have been proposed to address some shortcomings of the STL data format but their usage is still very limited.

STL files are generated through tessellation of accurate CAD models. The surfaces of the 3D models are approximated with triangular facets. Each triangle is defined independently by its three vertices and its outward normal vector. Two very important requirements must be followed during STL file generation [Jacob, 1996].

1. Data about triangle vertices must be stored in the file in an ordered fashion to identify interior or exterior surfaces. A clockwise vertex ordering defines an interior surface, and an anti-clockwise ordering the exterior surface. The right-hand rule (Figure 1.9) is applied.

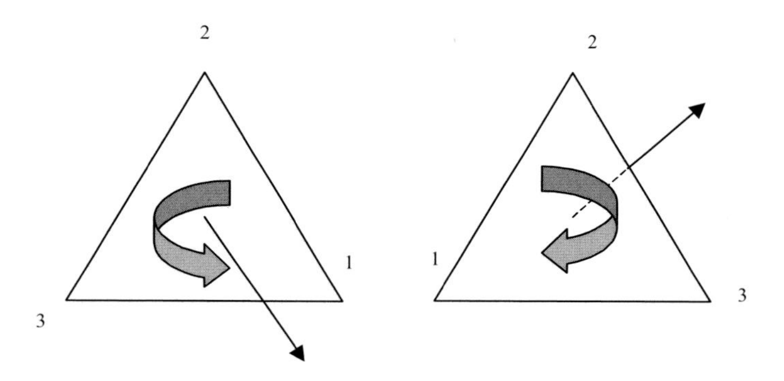

Figure 1.9 Right-hand rule

2. A triangle must share exactly two common vertices with each adjacent triangle (Figure 1.10). This is known as the vertex-to-vertex rule.

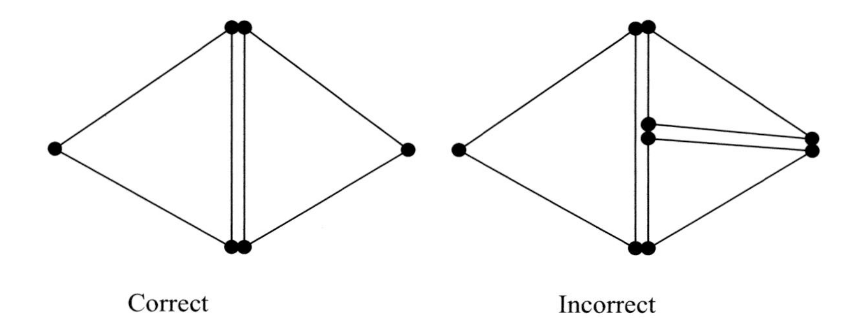

Figure 1.10 Vertex-to-vertex rule

Virtually every 3D modelling package provides an interface for exporting the internal CAD representation into the STL format.

The STL file can be stored in two formats, ASCII or binary. The binary files have a smaller size and therefore are preferred. Formally, the STL file format consists of a header and the number of triangles followed by the list of triangles [Jacobs, 1996]. 50 bytes are required to store one triangle in a binary format. Hence, the size of the binary file is equal to the number of triangles times 50 plus 84 bytes for the header and the triangle counter.

The use of other standard formats for product data exchange such as IGES (Initial Graphical Exchange Specification) [Reed et al., 1990], HPGL (Hewlett-Packard Graphics Language), STEP (STandard for Exchange of Product model data) [Owen, 1993] and VRML (Virtual Reality Modelling Language) [VRML, 2000] has been considered in place of STL but as problems remain these alternative formats are not widely accepted. The work on the development of new formats continues to address the growing requirements of RP applications for a more precise method of data representation.

1.3 RP Information Workflow

All RP systems have a common information workflow (Figure 1.11). The main stages in preparing and pre-processing data for automated fabrication of 3D objects are as follows:

1. *Data Creation*. A 3D CAD package or 2D scanning device can be employed to create geometric data. In both cases, the data must be represented in a model whose surfaces define a closed 3D volume without any holes, surfaces with zero-thickness or more than two surfaces meeting along common edges [Wozny, 1997]. Formally, the model is valid if the system can determine uniquely for each point in the 3D space whether it lies inside, on, or outside the object surfaces.

2. *Data Export*. The valid 3D model is exported from the CAD package in a neutral format, which in most cases is STL. Some CAD packages allow the size of the generated file to be controlled by increasing or reducing the model resolution.

3. *Data Validation and Repair*. The exported data is an approximation of the precise internal 3D model. During this approximation process the model surfaces are represented with simple geometrical entities in the form of triangles. Unfortunately, STL models created in this way can contain undesirable geometrical errors such as holes and overlapping areas along surface boundaries

[Wozny, 1997]. Therefore, the generated files have to be validated before being further processed. Some RP packages offer facilities for model repair, automatic and/or manual. These packages include software tools that evaluate the STL models and determine whether any triangles are missing. In case of errors, the gaps in the models are filled with new triangles.

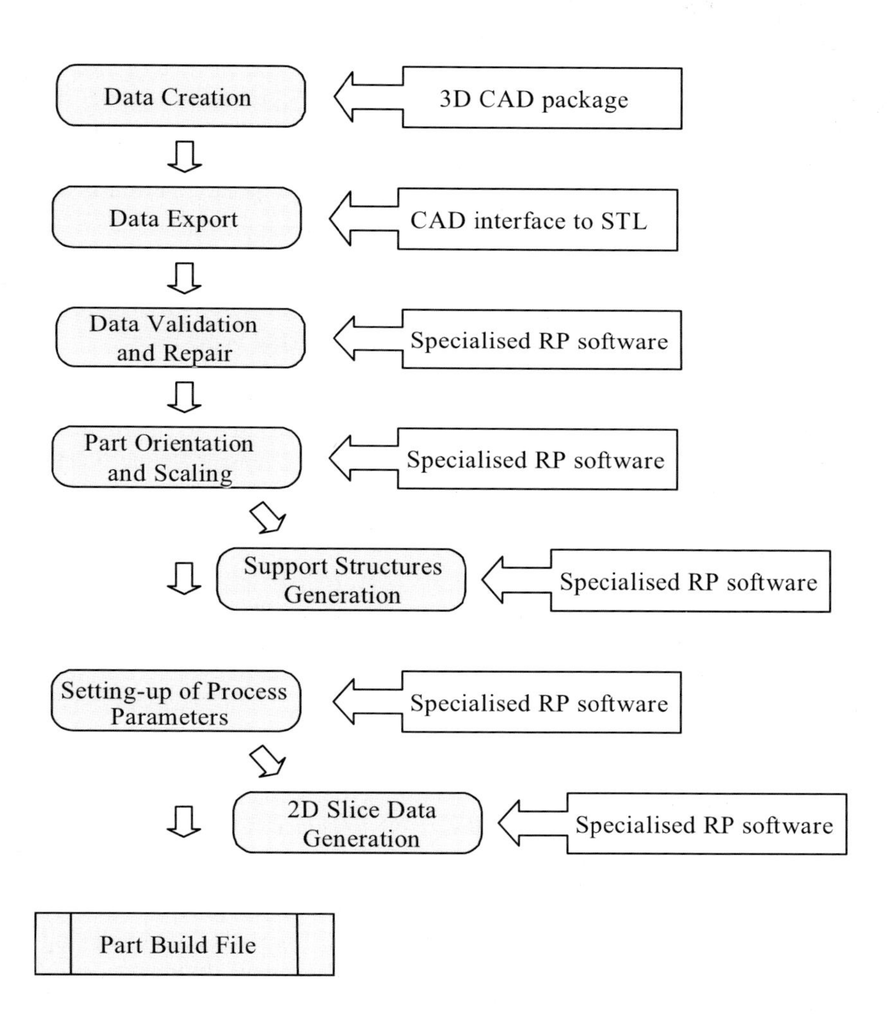

Figure 1.11 RP information workflow

4. *Part Orientation and Scaling*. RP systems build parts along the Z axis of their STL models. Through reorientation of the parts relative to the model coordinate systems, their accuracy, surface finish and build time can be optimised. Some RP systems allow several parts to be nested in the system chamber in order to be built simultaneously. In addition, the parts can be scaled to compensate for anticipated anomalies that might be introduced by downstream processes such as deformation, shrinkage, warpage and curling.

5. *Support Structures Generation*. Liquid-based RP processes require support structures to build overhanging areas of the parts. These structures are usually generated automatically employing specialised software tools. The areas requiring support structures can be minimised by appropriately selecting the part build direction.

6. *Setting-up of Process Parameters*. Process-related parameters are entered to specify the build style and desired system attributes. These parameters can be adjusted based on part requirements and the RP material being used.

7. *2D Slice Data Generation*. The STL file is sliced to produce successive cross-sectional layers. In each cross section, polylines are used to approximate the exterior and interior boundaries of the RP models. These polyline boundaries can be offset by a particular value to compensate for process errors. The slice data can be generated off-line for the entire model or on-line, one cross-section at a time during part building.

The process data generated following the stages outlined above is stored in a build file. This file contains all the information needed to guide material additive processes to build 3D objects.

1.4 Summary

RP technologies have emerged as a key element of TCE with their ability to shorten the product design and development process. This chapter has provided an overview of the chronological developments in the areas of RP. The role of 3D modelling as a gateway to RP has been discussed and the advantages and disadvantages of the available geometrical representation techniques outlined. Also, the existing formats for interfacing CAD and RP systems have been described together with the main stages in generating the necessary data to build 3D objects in layers.

References

Baese C (1904) **U.S. Patent** 774,549.

Beaman JJ (1997) Historical Perspective, Chapter 3 in **JTEC/WTEC Panel Report on Rapid Prototyping in Europe and Japan**, WTEC Hyper-Librarian (http://itri.loyola.edu/rp/toc.htm).

Blanther JE (1892) **U.S. Patent** 473,901.

Brite EuRam (1994) Common layer interface (CLI), Version 1.31. **Brite EuRam Project BE2578**, RPT – Development and Integration of Rapid prototyping Techniques for Automotive Industry.

DiMatteo PL (1976) Method of generating and constructing three-dimensional bodies, **U.S. Patent** 3,932,923.

Dolenc A and Malela I (1992) Leaf: a data exchange format for LMT processes, **Proc. 3th International Conference on Rapid Prototyping**, Dayton, USA, pp 4-12.

Gaskin TA (1973) **U.S. Patent** 3,751,827.

Herbert AJ (1982) Solid object generation. **J. Appl. Photo. Eng**., Vol. 8, 4, pp 185-88.

Jacobs PF (1996) Stereolithography and Other RP&M Techniques, **ASME Press**, New York.

Kai CC, Jacob GGK and Mei T (1997) Interface between CAD and Rapid Prototyping systems. Part 2: LMI – An Improved Interface, **Int J. Adv. Manuf. Technol.,** Vol. 13, pp 571-576.

Kodama H (1981) Automatic method for fabricating a three-dimensional plastic model with photo-hardening polymer. **Rev. Sci. Instrum**, pp 1770-1773.

Kunieda M and Nakagawa T (1984) Development of laminated drawing dies by laser cutting. **Bull. of JSPE**, pp 353-354.

Matsubara K (1974). **Japanese Kokai Patent Application**, Sho 51[1976]-10813.

Monteah FH (1924) **U.S. Patent** 1,516,199.

Morioka I (1935) **U.S. Patent** 2,015,457.

Morioka I (1944) **U.S. Patent** 2,350,796.

Munz OJ (1956) **U.S. Patent** 2,775,758.

Nakagawa T et al. (1979) Blanking tool by stacked bainite steel plates. **Press Technique**, pp 93-101.

Nakagawa T, Kunieda M and Liu S (1985) Laser cut sheet laminated forming dies by diffusion bonding, **Proc. 25th International MTDR Conference**, pp 505-510.

Owen J (1993) STEP An Introduction, **Information Geometers**.

Perera BV (1940) **U.S. Patent** 2,189,592.

Reed K, Harrvd D and Conroy W (1990) Initial Graphics Exchanges Specification (IGES) version 5.0, **CAD-CAM Data Exchange Technical Centre**.

Rock SJ and Wozny MJ (1991) A flexible file format for solid freeform fabrication, **Proceedings of Solid Freeform Fabrication Symposium**, Texas, USA, pp 155-160.

Swainson WK (1977) Method, medium and apparatus for producing three-dimensional figure product, **U.S. Patent** 4,041,476.

VRML Repository (2000) **Web3D Consortium**, http://www.web3d.org/vrml/vrml.htm.

Wohlers T (2000) Wohlers Report 2000: Executive Summary, **Time-Compression Technologies**, Vol. 8, 4, pp 29-31.

Wozny MJ (1997) CAD and Interfaces, Chapter 8 in **JTEC/WTEC Panel Report on Rapid Prototyping in Europe and Japan**, WTEC Hyper-Librarian (http://itri.loyola.edu/rp/toc.htm).

Zang EE (1940) **U.S. Patent** 3,137,080.

Zeid I (1991) CAD/CAM Theory and Practice, **McGraw-Hill**, Singapore.

Chapter 2 Rapid Prototyping Processes

This chapter presents a classification of existing physical Rapid Prototyping (RP) processes along with an outline of each method. For convenience, the term RP will hereafter refer only to physical RP.

2.1 Classification of Rapid Prototyping Processes

RP processes may be divided broadly into those involving the addition of material and those involving its removal. According to Kruth [Kruth, 1991], material accretion processes may be divided by the state of the prototype material before part formation, namely, liquid, powder or solid sheets. Liquid-based processes may entail the solidification of a resin on contact with a laser, the solidification of an electrosetting fluid, or the melting and subsequent solidification of the prototype material. Processes using powders aggregate them either with a laser or by the selective application of binding agents. Those processes which use solid sheets may be classified according to whether the sheets are bonded with a laser or with an adhesive. Figure 2.1 shows Kruth's classification which has been adapted to include new processes. In the following, RP processes are presented according to the arrangement shown in this figure.

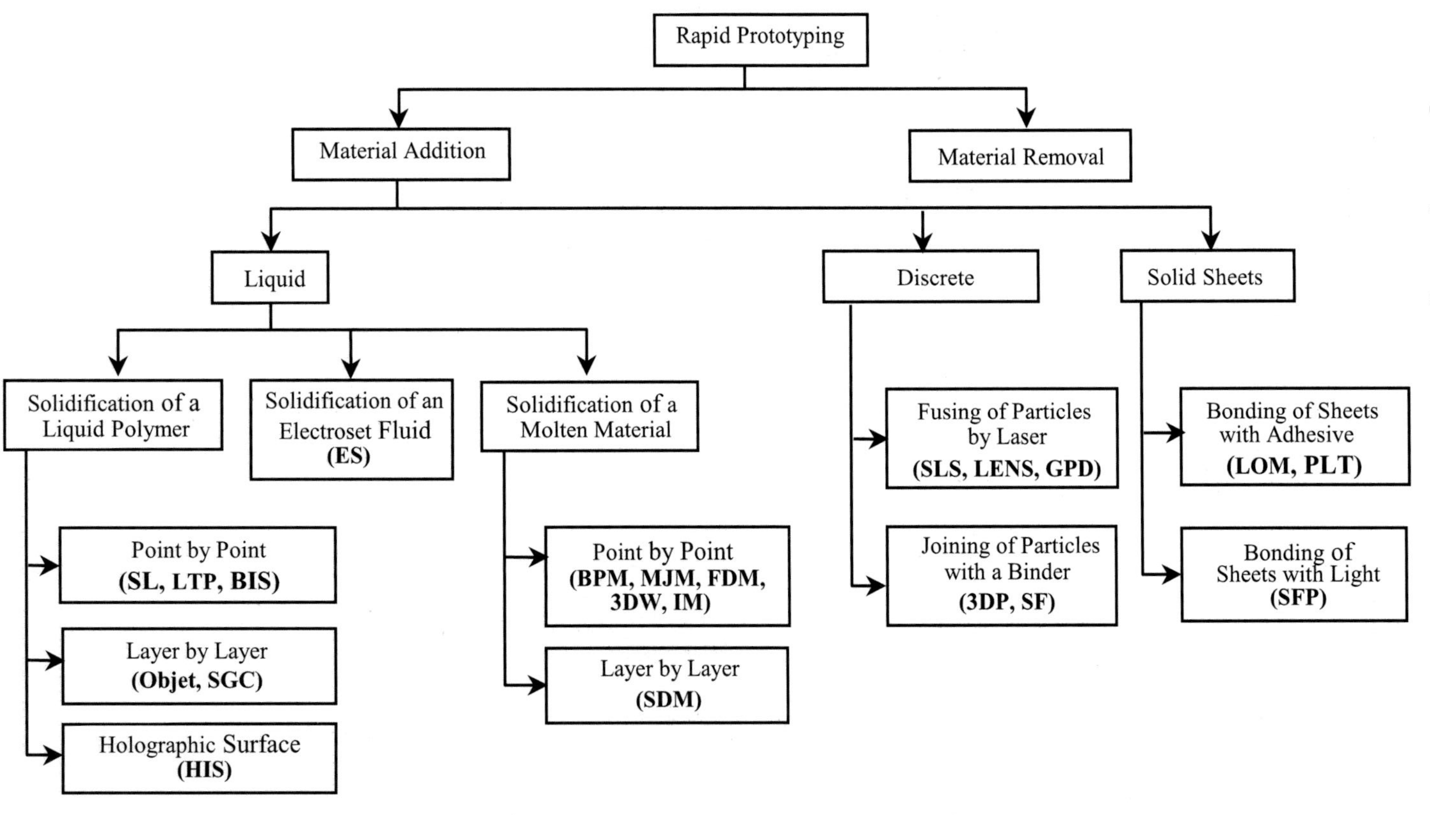

Figure 2.1 Classification of rapid prototyping methods (adapted from Kruth [Kruth, 1991])

2.2 Processes Involving a Liquid

2.2.1 Solidification of a Liquid Polymer

Of the six processes in this category, which all involve the solidification of a resin via electromagnetic radiation, three construct the part using points to build up the layers whilst the other three solidify entire layers or surfaces at once.

2.2.1.1 Stereolithography (SL)

The most popular among currently available RP processes is perhaps stereolithography (SL). This relies on a photosensitive liquid resin which forms a solid polymer when exposed to ultraviolet (UV) light. Due to the absorption and scattering of the beam, this reaction only takes place near the surface. This produces voxels (three-dimensional pixels), as shown in Figure 2.2, which are characterised by their horizontal line-width and vertical cure depth [Jacobs, 1992].

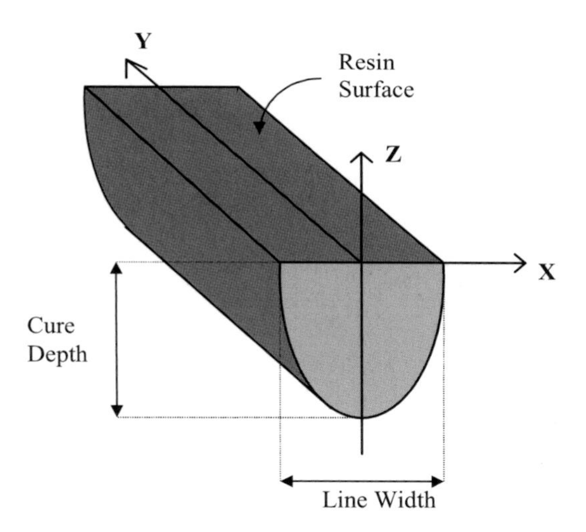

Figure 2.2 Single cured line of photopolymer [Jacobs, 1992]

A SL machine consists of a build platform (substrate) which is mounted in a vat of resin and a UV Helium-Cadmium or Argon ion laser (Figure 2.3). The first layer of the part is imaged on the resin surface by the laser using information obtained from the 3D solid CAD model. Once the contour of the layer has been scanned and the interior either hatched or solidly filled, the platform is next lowered to the base of the vat in order to coat the part thoroughly. It is then raised such that the solidified part is level with the surface and a blade wipes the resin so that exactly one layer-thickness remains above the part. The part is then lowered to one layer-thickness below the surface and left until the liquid has settled [Renap and Kruth, 1995]. This is done to ensure a flat and even surface and to inhibit bubble formation. The next layer may then be scanned.

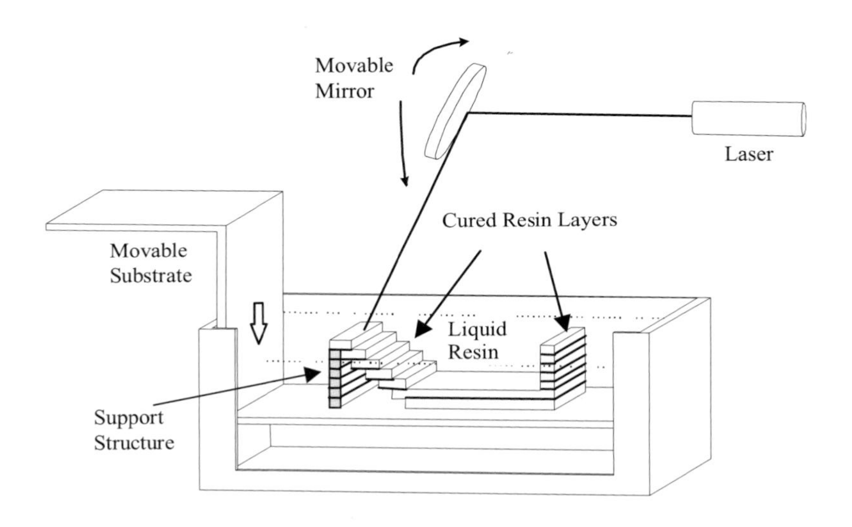

Figure 2.3 Stereolithography

All new SL machines now employ a method to apply the resin that is superior to the deep-dip process described above. Because of the high resin viscosity, after the deep dip and recoating, either too little or too much resin is left by the recoating blade, which affects part accuracy. The new method involves spreading resin on the part as the blade traverses the vat. Because the blade applies only the required amount of resin, good accuracy is achieved. This method also provides a smoother surface finish and reduces non-productive recoating time. Another important advantage is the elimination of 'trapped volumes'. A trapped volume is a volume of resin that cannot drain through the base of the part (Figure 2.4). The presence of a trapped volume in the deep-dip process affects part accuracy and may lead to delamination or collision of the blade and part because of a build up of unwanted polymerised resin at the surface of the trapped volume.

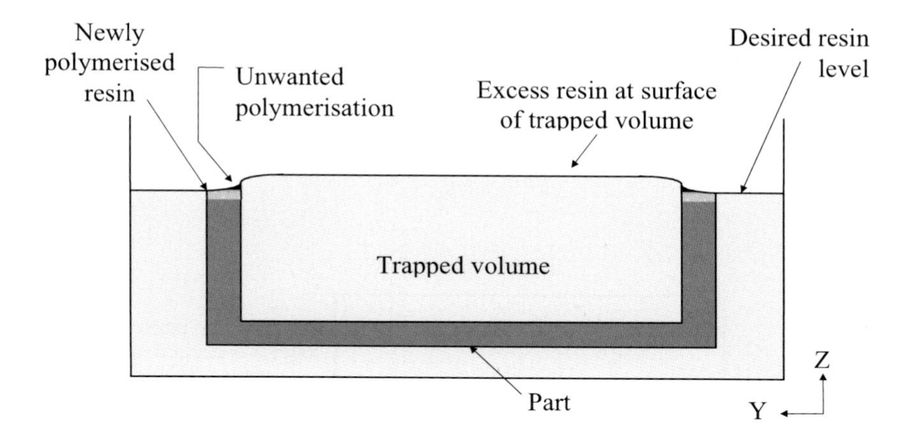

Figure 2.4 Trapped Volume in Stereolithography

Once the part is completed, it is removed from the vat and the excess resin drained. Due to the resin viscosity, this stage may take several hours. The 'green' part is then placed in an UV oven to be postcured. This ensures that no liquid or partially cured resin remains.

Further research is being actively conducted into materials and into the accuracy, warping and shrinkage of the parts. For more information about SL systems and their technical specifications, see Chapter 3.

2.2.1.2 Liquid Thermal Polymerisation (LTP)

This process is similar to SL except that the resin is thermosetting and an infrared laser is used to create the voxels. This difference means that the size of the voxels may be affected through heat dissipation, which can also cause unwanted distortion and shrinkage in the part. However, these problems are apparently no worse than those caused by SL and are controllable [Kruth, 1991]. The system is still being researched.

2.2.1.3 Beam Interference Solidification (BIS)

This process uses two laser beams mounted at right angles to each other which emit light at different frequencies to polymerise resin in a transparent vat (Figure 2.5). The first laser excites the liquid to a reversible metastable state and then the incidence of the second beam polymerises the excited resin.

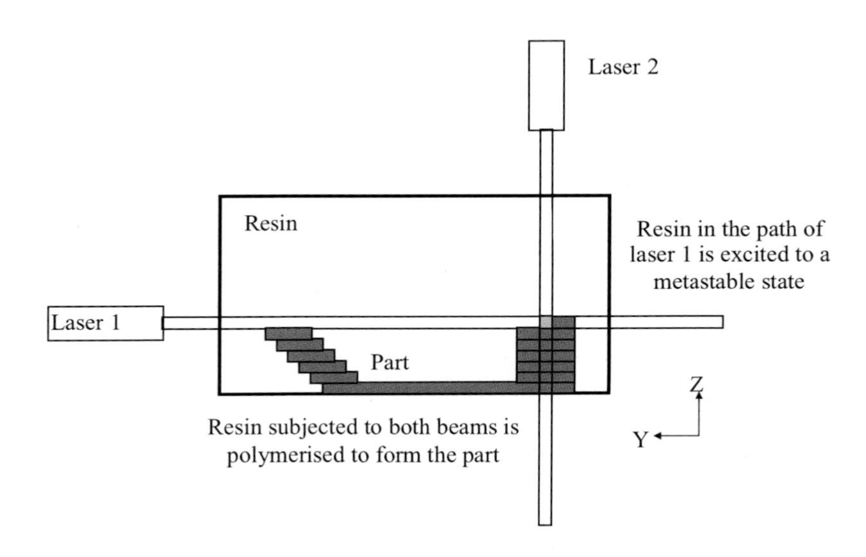

Figure 2.5 Beam Interference Solidification (Adapted from Kruth [Kruth, 1991])

To date, no commercial applications of this process exist because there are still technical difficulties to be solved:

- Shadows are cast from previously solidified sections.
- There is a problem with light absorption because the intensity of the lasers drops with depth.
- It is hard to intersect the laser beams due to diffraction variations in the resin caused by temperature gradients or solid sections [Kruth, 1991].

2.2.1.4 Objet Quadra Process

The Objet Quadra process employs 1536 nozzles to build parts by spreading layers of photo sensitive resin that are then cured, layer by layer, using two UV lights. The intensity of the lights and the exposure are controlled so that models produced by the system do not require post-curing. The Objet system prints with a resolution of 600 dpi, with a layer thickness of 20 μm. Only one photopolymer is currently available for building models but other materials are under development. To support overhanging areas and undercuts, Objet deposits a second material which can be separated from the model without leaving any contact points or blemishes [Objet, 2000]. For more information about the Objet system and its technical specifications, see Chapter 4.

2.2.1.5 Solid Ground Curing (SGC)

This system again utilizes photopolymerising resins and UV light (Figure 2.6). Data from the CAD model is used to produce a mask which is placed above the resin surface. The entire layer can then be illuminated with a powerful UV lamp. This means that the resin is fully cured and that no postcuring is necessary. Once the layer has been cured, the excess resin is wiped away and any spaces are filled with wax. The wax is cooled with a chill plate, milled flat and the wax chips removed. A new layer of resin is applied and the process is repeated.

The mask itself is a sheet of glass which is prepared whilst the current layer is being waxed, cooled and milled. The negative image of each subsequent layer is produced electrostatically on the glass and developed using a toner in a similar manner to laser printing.

Because wax is used to fill the gaps in the cured resin, no further supports need to be added by the interface software. The wax supports any overhangs in the design and anchors any discrete protrusions which may be drawn on a layer. It also theoretically reduces distortion due to warping and curling since the part is surrounded and means that the machines do not need to be vibration proofed as the part cannot move in the vat [Kruth, 1991; Au and Wright, 1993]. Builds may also be suspended to allow other, more urgent parts to be made [Cubital, 1996]. An advantage of this system is that the entire layer is solidified at once, reducing the part creation time, especially for multi-part builds. Parts may also be nested to utilise the build volume fully. All the resin within a layer is completely cured by this method, and so no postcuring is required, parts may be more durable than the hatched prototypes created using other processes and operators need not handle partially cured, toxic resin [Cubital, 1996]. The wax may be removed automatically in a special machine.

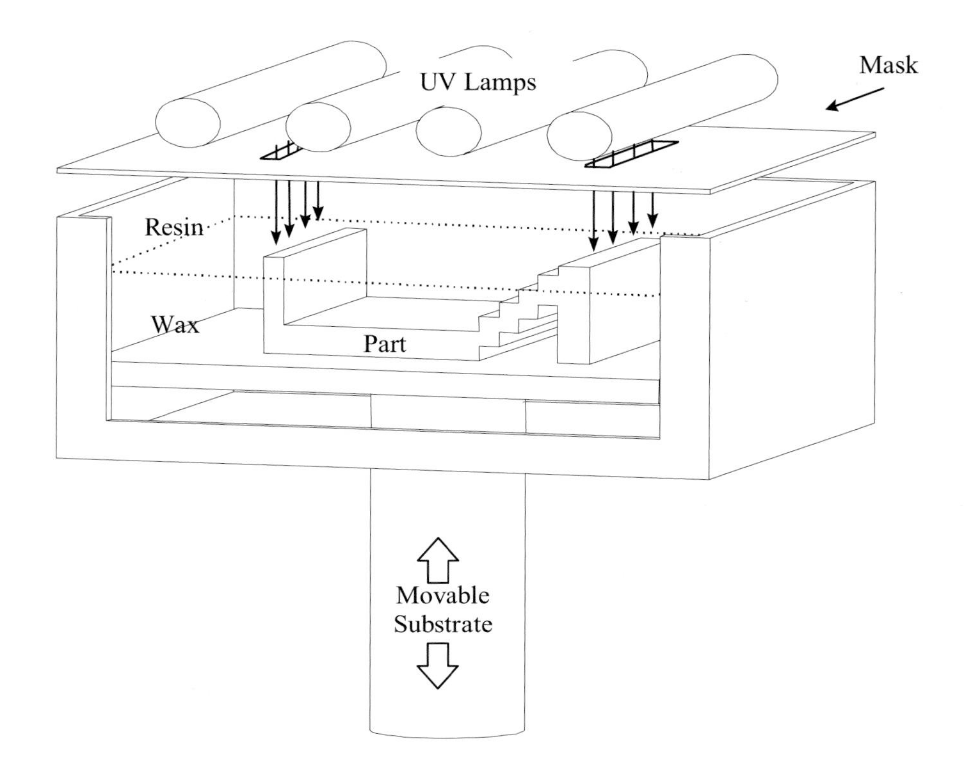

Figure 2.6 Solid Ground Curing (SGC)

The disadvantages of this system are that it is noisy, large and heavy and needs to be constantly manned. It wastes a large amount of wax which cannot be recycled and is also prone to breakdowns [Waterman and Dickens, 1994; Jacobs, 1996]. The mask is produced by raster scanning the image [Cubital, 1996] which may cause steps in the x-y plane, affecting accuracy. The resin models produced using SGC are solid and so cannot be used for later investment casting since the coefficient of thermal expansion of the resin is an order of magnitude greater than that of the ceramic system so the ceramic moulds will crack when the sacrificial part is burnt out [Waterman and Dickens, 1994].

For more information about SGC systems and their technical specifications, see Chapter 3.

2.2.1.6 Holographic Interference Solidification (HIS)

A holographic image is projected into the resin causing an entire surface to solidify. Data is still obtained from the CAD model, although not as slices. The build space is 300x300x300 mm [Kruth, 1991]. There are no commercial systems available yet.

2.2.2 Solidification of an Electroset Fluid: Electrosetting (ES)

Electrodes are printed onto a conductive material such as aluminium. Once all the layers have been printed, they are stacked, immersed in a bath of electrosetting fluid and energised. The fluid which is between the electrodes then solidifies to form the part. Once the composite has been removed and drained, the unwanted aluminium may be trimmed from the part.

Advantages of this technology are that the part density, compressibility, hardness and adhesion may be controlled by ajusting the voltage and current applied to the aluminium. Parts may be made from silicon rubber, polyester, polyurethane or epoxy. The hardware for such a system may be bought off the shelf at a low cost. The software for the system is still being developed [Anon, 1993].

2.2.3 Solidification of Molten Material

There are six processes which involve the melting and subsequent solidification of the part material. Of these, the first five deposit the material at discrete points whilst the sixth manufactures whole layers at once.

2.2.3.1 Ballistic Particle Manufacture (BPM)

A stream of molten material is ejected from a nozzle. It separates into droplets which hit the substrate and immediately cold weld to form the part (Figure 2.7). If the substrate is rough, thermal contact between it and the part is increased which reduces stresses within the part [Anon, 1995].

The stream may be a drop-on-demand system or a continuous jet. When a continuous jet is adopted, it is ejected through a nozzle which is excited by a piezoelectric transducer at a frequency of about 60 Hz [Sachs et al., 1992]. To avoid melting the transducer, it is located at a distance from the nozzle. Although a capillary stream will naturally decompose into droplets [Rayleigh, 1878], the disturbance at the nozzle forces the production of a stream of small, regular droplets with uniform spacing and distance. Using a low-frequency carrier wave modulated by a higher-frequency disturbance, tailor-made streams have been produced where the user is able to specify larger droplet separations than would otherwise be obtainable with just a single frequency. Regular streams have also been produced consisting of a few small droplets close together followed by larger, more widely spaced droplets [Orme et al., 1994]. This should allow more time for the nozzle to move to a new position or for the droplets to solidify if necessary.

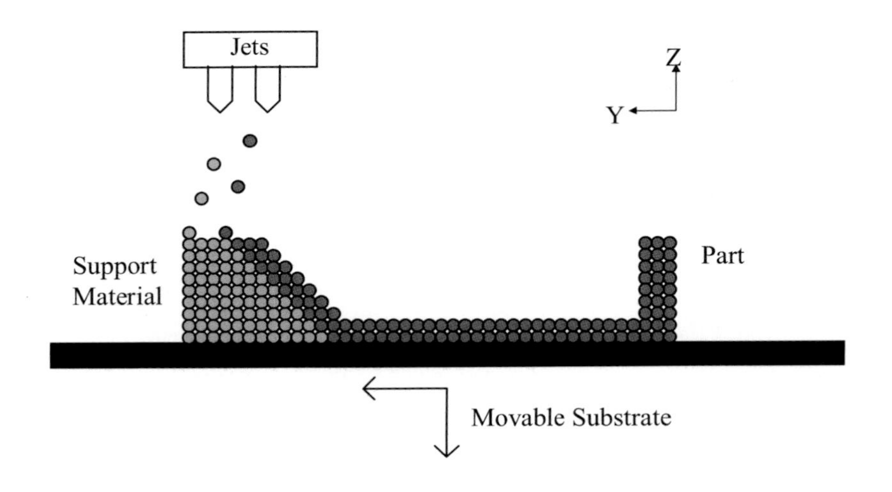

Figure 2.7 Ballistic Particle Manufacture

Parameters that will affect the eventual part characteristics are the temperature and velocity of the droplets and the charge that they carry. The charge is acquired electrostatically when the stream is ejected and is used for the accurate placement of the material. Since the maximum charge that may be held by a drop is limited, the maximum deflection of such a drop is also limited and the substrate or the jet must therefore be movable in order to produce a large enough build area. The temperature will control the speed at which the molten material solidifies. If the droplets are too cold, they will solidify mid-flight and will therefore not weld to the part. If they are too hot, the part will lose its shape. The deformation and placement accuracy of a droplet depend on its velocity. If it is moving too slowly, the placement accuracy will be poor; if it moves too quickly, the droplet will be highly deformed on impact [Anon, 1995].

2.2.3.2 Multi Jet Modelling (MJM)

A MJM machine builds models using a technique similar to inkjet or phase-change printing but applied in three dimensions [3D Systems, 1998]. A "print" head comprising 352 jets oriented in a linear array builds models in successive layers, each individual jet depositing a specially developed thermo-polymer material only where necessary (Figure 2.8). The layer thickness is 40 μm. The MJM head shuttles back and forth along the X axis like a line printer. If the part is wider than the MJM head, the platform repositions (Y-axis) to continue building the layer. When the layer is complete, the platform is moved away from the head (Z-axis) which begins to create the next layer. When the build is completed, support structures are brushed off to finish the model.

For more information about the MJM system and its technical specifications, see Chapter 4.

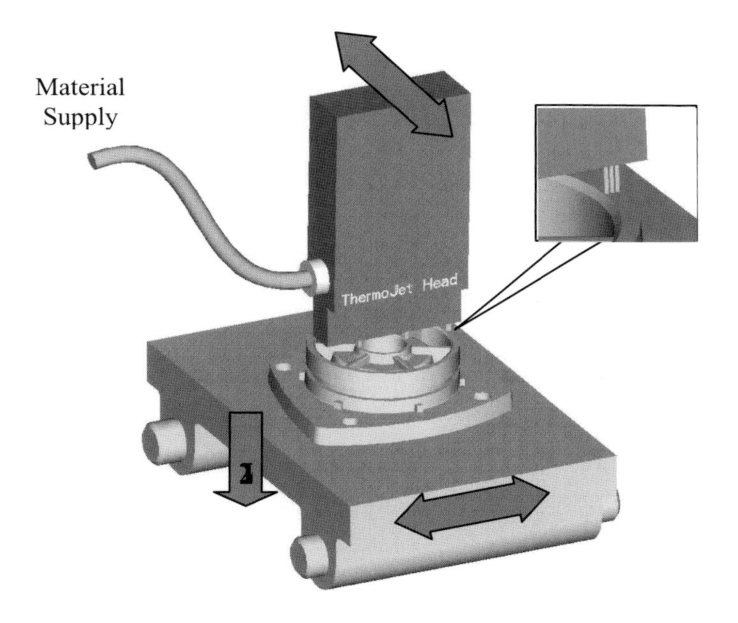

Figure 2.8* Multi Jet Modelling Head

2.2.3.3 Fused Deposition Modelling (FDM)

A FDM machine consists of a movable head which deposits a thread of molten material onto a substrate. The build material is heated to 0.5°C above its melting point so that it solidifies about 0.1 s after extrusion and cold welds to the previous layers (Figure 2.9). Factors to be taken into consideration are the necessity for a steady nozzle speed and material extrusion rate, the addition of a support structure for overhanging parts, and the speed of the head which affects the overall layer thickness [Au and Wright, 1993; Stratasys, 1991].

More recent FDM systems include two nozzles, one for the part material and one for the support material. The latter is cheaper and breaks away from the prototype without impairing its surface. It is also possible to create horizontal supports to minimise material usage and build time [Stratasys, 1991; Stratasys, 1996]. More detailed information about FDM systems can be found in Chapter 3.

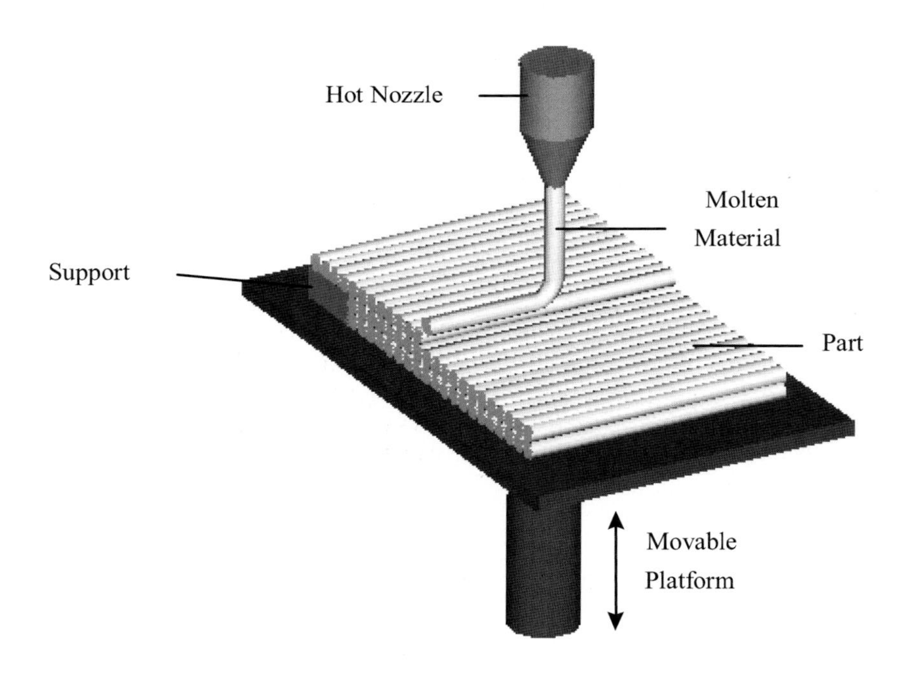

Figure 2.9 The Fused Deposition Modelling process

2.2.3.4 Three-Dimensional Welding (3DW)

This system uses an arc-welding robot to deposit material on a platform as simple shapes which may then be built into more complex structures. Unlike most RP processes, the prototypes are not built using sliced CAD files.

Several problems still remain to be solved. Since there is no feedback, heat build-up during manufacture can cause the prototypes to melt and because the layers do not form a smooth surface the torch may hit the part [Dickens, 1995; Dickens et al., 1992]. It is also not known whether complex structures can be built. Some method needs to be found to generate the robot program directly from the CAD file. The orientation of each section to be built should be generated as well as the order in which the sections are to be assembled.

Another system which is being researched deposits the weld material in layers. Feedback control is established by the use of thermocouples which monitor the temperature and operate an on-line water cooling system. A grit blasting nozzle minimises the oxidisation of the part and a suction pump and vacuum nozzle remove excess water vapours and grit [Anon, 1993].

2.2.3.5 Shape Deposition Manufacturing (SDM)

Still experimental this layer-by-layer process involves spraying molten metal in near net shape onto a substrate, then removing unwanted material via NC operations. Support material is added in the same way either before or after the prototype material depending on whether the layer contains undercut features (Figure 2.10). The added material bolsters subsequent layers. If the layer is complex, support material may need to be added both before and after the prototype material. Each layer is then shot-peened to remove residual stresses. The prototype is transferred from station to station using a robotised pallet system which can position the workpiece to within an accuracy of ±5 μm. Droplets of 1-3 mm diameter are deposited at a rate of 1-5 droplets per second.

To date, stainless steel parts supported with copper have been produced. The copper may then be removed by immersion in nitric acid. These prototypes have the same structure as cast or welded parts and the accuracy of NC milled components.

Multiple materials may be employed and components can be embedded in the structure. As yet, no temperature control system for the substrate has been implemented, and the temperature, size and trajectory of the droplets are also not controlled [Merz et al., 1994].

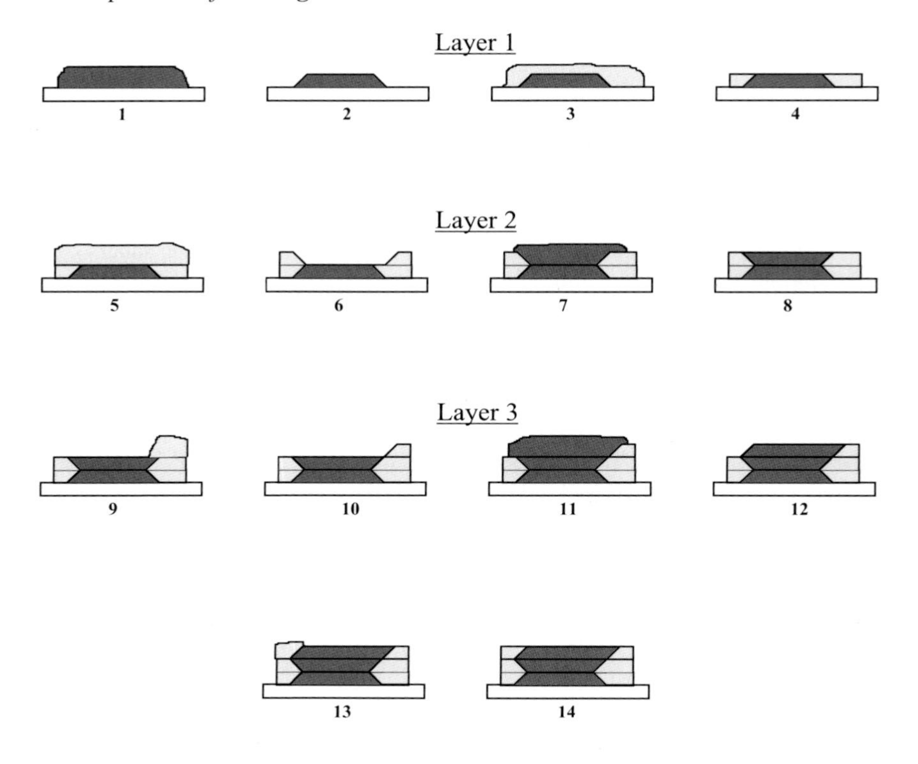

Figure 2.10 Shape Deposition Manufacturing. The construction of the first 3 layers of a part is shown (from [Merz et al., 1994])

1. Layer 1: part material is added
2. Layer 1: part material is milled
3. Layer 1: support material is added
4. Layer 1: support material is milled
5. Layer 2: support material is added
6. Layer 2: support material is milled
7. Layer 2: part material is added
8. Layer 2: part material is milled
9. Layer 3: support material is added
10. Layer 3: support material is milled
11. Layer 3: part material is added
12. Layer 3: part material is milled
13. Layer 3: support material is added
14. Layer 3: support material is milled

2.3 Processes Involving Discrete Particles

These processes build the part by joining powder grains together using either a laser or a separate binding material.

2.3.1 Fusing of Particles by Laser

Selective Laser Sintering (SLS) and Laser Engineering Net Shaping (LENS^{TM}) are the main processes in this category. In the Gas Phase Deposition (GPD) process, discrete grains are formed as a result of interaction between a reactive gas and a laser and the laser also fixes the grains with respect to the part.

2.3.1.1 Selective Laser Sintering (SLS)

SLS uses a fine powder which is heated with a CO_2 laser so that the surface tension of the particles is overcome and they fuse together. Before the powder is sintered, the entire bed is heated to just below the melting point of the material in order to minimize thermal distortion and facilitate fusion to the previous layer [Klocke et al., 1995]. The laser is modulated such that only those grains which are in direct contact with the beam are affected. A layer is drawn on the powder bed using the laser to sinter the material. The bed is then lowered and the powder-feed chamber raised so that a covering of powder can be spread evenly over the build area by a counter-rotating roller. The sintered material forms the part whilst the unsintered powder remains in place to support the structure and may be cleaned away and recycled once the build is complete (Figure 2.11).

More detailed information about SLS systems and their technical specifications can be found in Chapter 3.

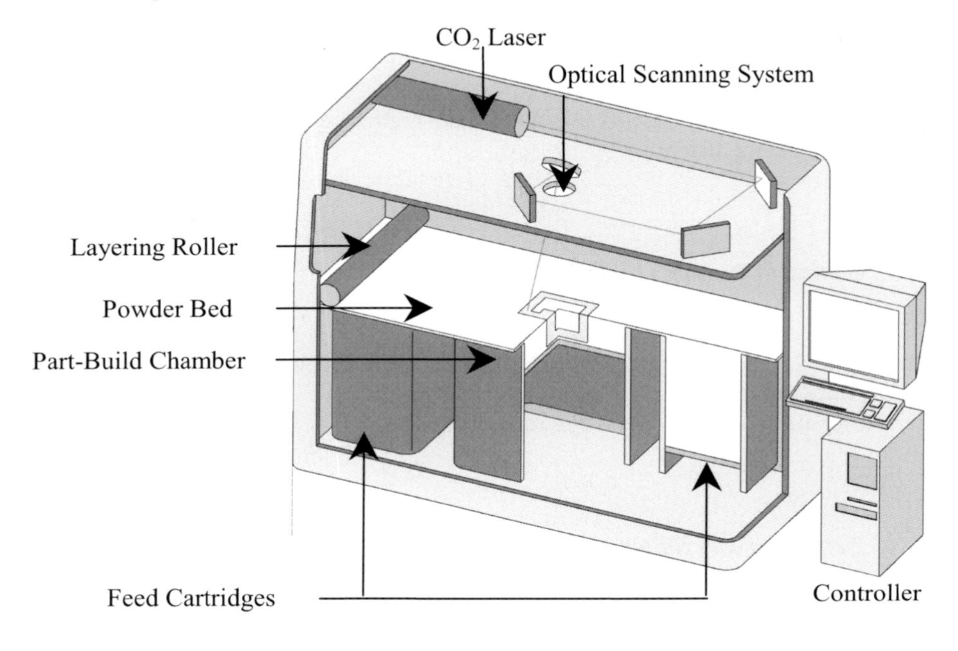

Figure 2.11 Selective Laser Sintering

2.3.1.2 Laser Engineering Net Shaping (LENS™)

The LENS process involves feeding powder through a nozzle onto the part bed whilst simultaneously fusing it with a laser (Figure 2.12) [Optomec, 2000]. The powder nozzle may be on one side of the bed or coaxial with the laser beam. If it is to a side, a constant orientation to the part creation direction must be maintained to prevent solidified sections from shadowing areas to be built. When the powder feeder is coaxial, there may be inaccuracies in the geometry of the part and the layer thickness if the beam and the powder feeder move out of alignment.

Because the stream of powder is heated by the laser, fusion to the previous layer is facilitated. However, this also leads to thermal distortion of the part. It is necessary to cool the part when it becomes too hot in order to prevent distortions in the final piece. An alternative would be to add a temperature control system. The minimum wall thickness depends on the feed rate, the width of the particle stream and spot size, speed and power of the laser [Klocke, 1995]. More detailed information about LENS systems can be found in Chapter 3.

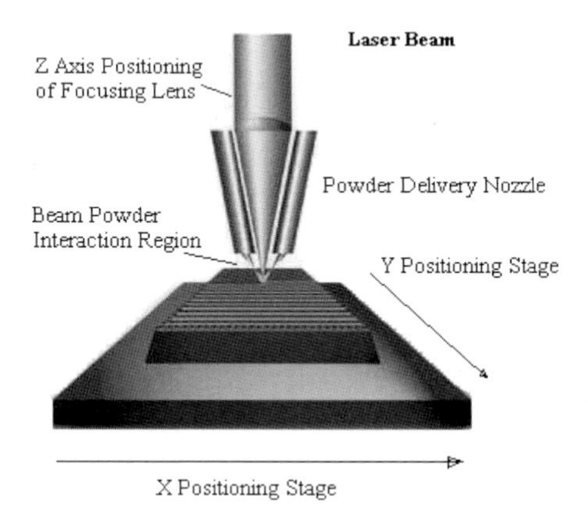

Figure 2.12* LENS™ process (Courtesy of Optomec Design Co)

2.3.1.3 Gas Phase Deposition (GPD)

In this process, the molecules of a reactive gas are decomposed using a laser to generate a solid. The resulting solid then adheres to the substrate to form the part (Figure 2.13). Three slightly different methods of constructing the part are currently being researched.

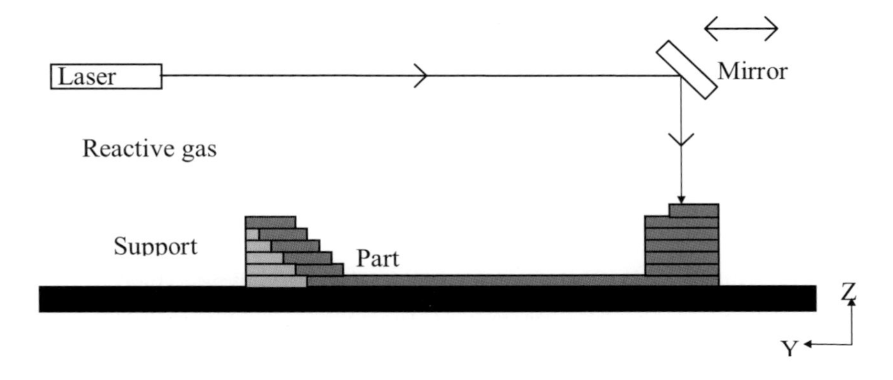

Figure 2.13 Gas Phase Deposition

With the first method, called SALD (Selective Area Laser Deposition), the solid component of the decomposed gas is all that is used to form the part. It is possible to construct parts made from carbon, silicon, carbides and silicon nitrides in this way. The second method, SALDVI (Selective Area Laser Deposition Vapour Infiltration), spreads a thin covering of powder for each layer and the decomposed solids fill in the spaces between the grains. With the third method, SLRS (Selective Laser Reactive Sintering), the laser initiates a reaction between the gas and the layer of powder to form a solid part of silicon carbide or silicon nitride [Dickens, 1995; Laboratory for Freeform Fabrication, 1996].

2.3.2 Joining of Particles with a Binder

2.3.2.1 Three-Dimensional Printing (3DP)

Layers of powder are applied to a substrate then selectively joined using a binder sprayed through a nozzle (Figure 2.14). In order to avoid excessive disturbance of the powder when it is hit by the binder, it is necessary to stabilise it first by making it moist with water droplets [Sachs et al., 1993]. Once the build is completed, the excess powder, which was supporting the model, is removed leaving the fabricated part. Since there is no state change involved in this process, distortion is reduced [Sachs et al., 1993].

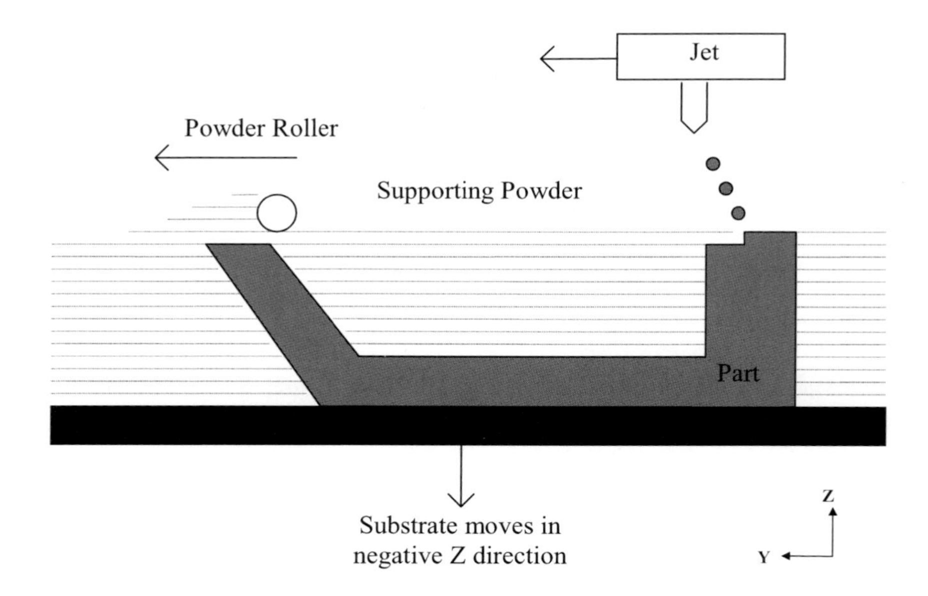

Figure 2.14 Three-Dimensional Printing

Parts made using this process do not require supports to brace overhanging features. They do however need to include a hole so that excess powder can be removed [Sachs et al., 1993]. The disadvantages of this technology are that the final parts may be fragile and porous, and the excess powder may be difficult to remove from any cavities. A further drawback is that the layers are raster-scanned by the printhead which leads to a stair-stepping effect in the X-Y plane as well as in the build direction [Jacobs, 1996].

More detailed information about 3DP systems can be found in Chapter 4.

2.3.2.2 Spatial Forming (SF)

This technology is being developed for prototyping specialised medical equipment in metal. It is designed to produce high precision parts within a small build envelope of 2 x 2 x 300 mm. A negative of each layer is printed onto a ceramic substrate with a ceramic pigmented organic ‘ink’. The layer is then cured with UV light and the process repeated. After approximately 30 layers, the positive space left by the printing, which corresponds to the part cross section, is filled using another ‘ink’ which contains metal particles. This is then cured and milled flat. The process continues until the whole part is finished. Once the prototype is completely built, it is heated in a nitrogen atmosphere to remove the binders in both the positive and negative ‘inks’ and to sinter the metal particles. The ceramic negative can then be removed in an ultrasonic bath to reveal the final piece, which is infiltrated with liquid metal to produce the metal prototype.

The sintering process causes shrinkage of up to 20% in all directions which needs to be taken into account when designing the part. Further research includes optimising the binder removal process and automating the addition of the positive material and the subsequent milling operation [Taylor, 1995].

A prototype of this system is currently being employed to construct pre-assembled microstructures for medical purposes. To date, no commercial system is available and only extruded parts with a constant cross-section can be produced. In theory, however, completely arbitrary geometries should be feasible.

2.4 Processes Involving Solid Sheets

There are three different processes that use foils to form the part. Laminated Object Manufacture (LOM) and Paper Lamination Technology (PLT) bond the different sheets with an adhesive and then cut the part contour using a laser and a computerised knife, respectively. Solid Foil Polymerisation (SFP) bonds foils of a polymeric material by curing them with UV light.

2.4.1 Laminated Object Manufacturing (LOM)

The build material is applied to the part from a roll, then bonded to the previous layers using a hot roller which activates a heat-sensitive adhesive [Helisys, 2000]. The contour of each layer is cut with a laser that is carefully modulated to penetrate to the exact depth of one layer. Unwanted material is trimmed into rectangles to facilitate its later removal but remains in place during the build to act as supports (Figure 2.15). The sheet of material used is wider than the build area so that, once the part cross-section has been cut, the edges of the sheet remain intact. This means that, after the layer has been completed and the build platform lowered, the roll of material can be advanced by winding this excess onto a second roller until a fresh area of the sheet lies over the part. The whole process can then be repeated.

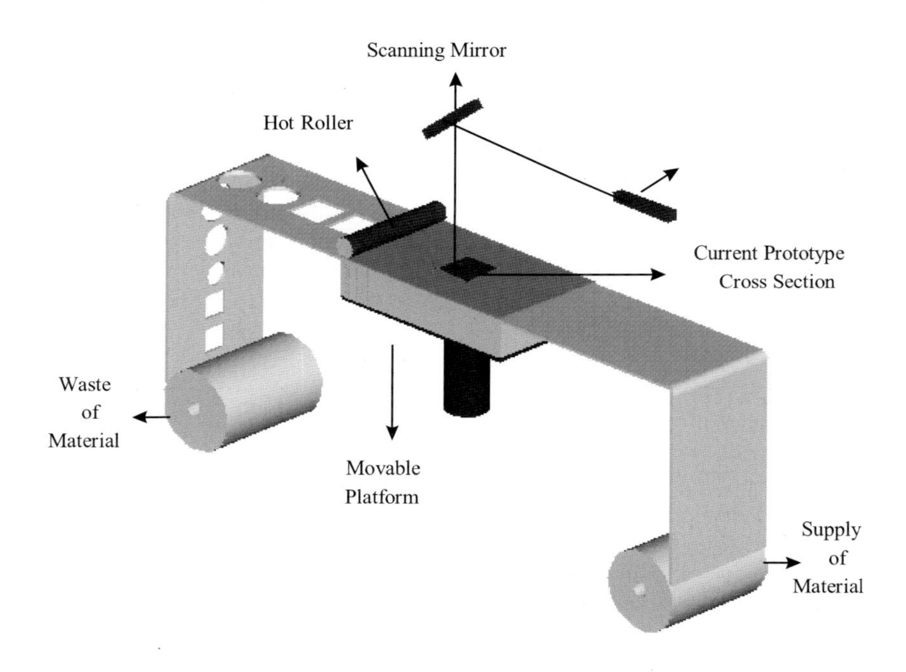

Figure 2.15 Laminated Object Manufacturing.

The system employs a 25 or 50 Watt CO_2 laser to cut the material. Smaller hatches must be used on up and down-facing surfaces to facilitate the removal of waste material which otherwise would be bonded to the part. It may also be necessary to stop the build to excavate paper from hard-to-access places. Once the parts are completed, they should be sealed with a urethane lacquer, silicone fluid or epoxy resin spray to prevent later distortion of the paper prototype through water absorption. The height is measured and the cross-sections are calculated in real time to correct for any errors in the build direction [Jacobs, 1996].

Advantages of LOM include the wide range of relatively cheap materials available - parts may be made using paper for example, or from more expensive materials such as plastic or fibre-reinforced glass ceramic. The parts may be large compared with those produced by other RP methods. Since they have the appearance of wooden pieces when finished, they are popular with model makers. Speed is another strong feature of LOM. As only the outlines of the parts need to be traced, this method is about 5 - 10 times faster than other processes [Helisys, 1997].

A drawback is the need to prise the finished parts off the build platform which adversely affects their surface finish. It is also hard to make hollow parts due to the difficulty in removing the core and there are serious problems with undercuts and re-entrant features. Other disadvantages of this technology are that there is a large amount of scrap, the machine must be constantly manned, parts need to be hand finished and the shear strength of the part is adversely affected by the layering of adhesive and foil [Waterman and Dickens, 1994; Crump, 1991]. Because the laser cuts through the material, there is a fire hazard which means that the machines need to be fitted with inert gas extinguishers. The drops of molten material (dross) which form during the cutting process also need to be removed [Kruth, 1991].

For more information about the LOM process and its applications, see Chapter 3.

2.4.2 Paper Lamination Technology (PLT)

In RP terms, the PLT process is very similar to LOM. The main differences between the LOM and PLT processes are in the material used and the methods employed for cutting the contours of the part cross-sections, which are a CO_2 laser and a computerised knife, respectively. The PLT process prints the cross section of the part onto a sheet of paper, which is then applied to the work-in-progress and bonded using a hot roller [Yuasa, 1997]. Next, a computer-driven knife is used to cut the outline of the part and cross-hatch the waste material. This process is repeated until the part is finished, when the excess material may be peeled away from the model, which can be then sealed with epoxy resin. More detailed information about PLT systems can be found in Chapter 3.

2.4.3 Solid Foil Polymerisation (SFP)

In SFP, the part is built up using semi-polymerised foils. On exposure to UV light, the foil solidifies and bonds to the previous layer. It also becomes insoluble. Once the cross-section has been illuminated, a new foil can be applied. The areas of foil which do not constitute the eventual part are used to support it during the build process, but remain soluble and so are easy to remove. Once the part is complete, the non-bonded pieces can be dissolved to leave the finished part [Kruth, 1991; Corbel et al., 1994]. No commercial systems are available yet.

2.5 Summary

This chapter has described the technologies currently available for rapidly building physical prototypes. The chapter has provided a classification of existing RP technologies followed by an outline of each method.

References

Anon (1993) State of the Art Review-93-01, **MTIAC**, 10 West 35 Street, Chicago, IL 60616, USA.

Anon (1995) Manufacturing Parts Drop By Drop, **Compressed Air**, March, Vol. 100, 2, pp 38-44.

Au S and Wright PK (1993) A Comparative Study of Rapid Prototyping Technology, **Proceedings ASME Winter Conference**, New Orleans, November, Vol. 66, pp 73-82.

Corbel S, Allanic AL, Schaeffer P and Andre JC (1994) Computer-Aided Manufacture of Three-Dimensional Objects by Laser Space-Resolved Photopolymerization, **Journal of Intelligent and Robotic Systems**, Vol. 9, pp 310-312.

Crump SS (1991) Fast, Precise, Safe Prototypes with FDM, **ASME Annual Winter Conference**, Atlanta, December, Vol. 50, pp 53-60.

Cubital Ltd. (1996) Advantages of the Solider System, **Cubital Ltd.**, 13 Hasadna St., PO Box 2375, Industrial Zone North, Raanana, 43650 Israel.

Dickens PM, Pridham MS, Cobb RC and Gibson I (1992) Rapid Prototyping Using 3-D Welding, **Proceedings of the 3rd Symposium on Solid Freeform Fabrication**, Austin, Texas, September, pp 280-290.

Dickens PM (1995) Research Developments in Rapid Prototyping, **Proceedings IMechE, Journal of Engineering Manufacture, Part B**, Vol. 209, pp 261-266.

Helisys Inc. (1997) 2030H System, **Helisys Inc.**, 24015 Garnier Street, Torrance, CA 90505-5319.

Helisys Web page (2000) **Helisys, Inc**., 24015 Garnier Street, Torrance, California 90505-5319, USA, http://helisys.com/.

Jacobs PF (1992) Fundamentals of Stereolithography, **First European Conference on Rapid Prototyping**, Nottingham, July, pp 1-17

Jacobs PF (1996) Stereolithography and Other RP&M Techniques, **ASME Press**, New York.

Klocke F, Celiker T and Song Y-A (1995) Rapid Metal Tooling, **Rapid Prototyping Journal**, Vol. 1, 3, pp 32-42

Kruth JP (1991) Material Incress Manufacturing by Rapid Prototyping Technologies, **CIRP Annals**, Vol. 40, 2, pp 603-614.

Laboratory for Freeform Fabrication (1996) **Web pages of University of Texas at Austin**, Texas, USA.

Merz R, Prinz FB, Ramaswami K, Terk M and Weiss LF (1994) Shape Deposition Manufacturing, **Proceedings of the 5th Symposium on Solid Freeform Fabrication**, Aug. 8-10, Austin, Texas, pp 1-8.

Objet Web page (2000) **Objet Geometries Ltd.** Rehovot, Israel, http://clients.tia.co.il/objet/inner/products.html.

Optomec Web page (2000) **Optomec Design Company**, 2701-D Pan American Freeway - Albuquerque, New Mexico - 87107, USA, http://www.optomec.com/.

Orme M, Willis K and Courter J (1994) The Development of Rapid Prototyping of Metallic Components via Ultra Uniform Droplet Deposition, **Proceedings of the 5th International Conference on Rapid Prototyping**, Dayton, Ohio, June 12-15, pp 27-37.

Rayleigh (1878) On the Instability of Jets, **Proceedings of the London Mathematical Society**, Vol. 10, Part 4, pp 4-13.

Renap K and Kruth JP (1995) Recoating Issues in Stereolithography, **Rapid Prototyping Journal**, Vol. 1, 3, pp 4-16.

Sachs E, Cima M, Williams P, Brancazio D and Cornie J (1992) Three Dimensional Printing: Rapid Tooling and Prototyping Directly from a CAD Model, **Transactions of ASME: Journal of Engineering for Industry**, November, Vol. 114, pp 481-488.

Sachs E, Cornie J, Brancazio D, Bredt J, Curodeau A, Fan T, Khanuja S, Lauder A, Lee J and Michaels S (1993) Three Dimensional Printing: the Physics and Implications of Additive Manufacturing, **CIRP Annals**, Vol. 42, 1, pp 257-260.

Stratasys Inc. (1991) Fused Deposition Modelling for Fast, Safe Plastic Models, **12th Annual Conference on Computer Graphics**, Chicago, April, pp 326-332.

Stratasys Inc. (1996) FDM-1650, **Stratasys Inc.**, 14950 Martin Drive, Eden Prairie, Minneapolis 55344-2020, USA.

Taylor CS, Cherkas P, Hampton H, Frantzen JJ, Shah BO, Tiffany WB, Nanis L, Booker P, Sabhieh A and Hansen R (1995) Spatial Forming, a Three-Dimensional Printing Process, **Proceedings IEEE Micro Electro Mechanical Systems Conference**, Amsterdam, January, pp 203-208.

Waterman NA & Dickens P (1994) Rapid Product Development in the USA, Europe and Japan, **World Class Design To Manufacture**, Vol. 1, 3, pp 27-36.

YUASA Warwick machinery Ltd. (1997), Private communication, **YUASA Warwick machinery Ltd.**, Rothwell Road, Wedgenock Ind. Est., Warwick CV24 5PY, UK .

3D Systems Press Release (1998) ThermoJet, **3D Systems**, Worldwide Corporation HQ, 26081 Avenue Hall, Valencia, California, USA.

Chapter 3 Technical Characteristics and Technological Capabilities of Rapid Prototyping Systems

This chapter gives a technical overview of commercially available RP systems. In particular, the chapter examines the technical characteristics and application areas of systems based on the following RP processes: stereolithography, solid ground curing, fused deposition modelling, selective laser sintering, laminated object manufacturing and laser engineering net shaping. The main strengths and weaknesses of these systems are also reviewed. In addition, the chapter gives examples of different RP applications in which these commercially available systems have been successfully employed.

3.1 Stereolithography Apparatus (3D Systems)

The Stereolithography Apparatus (SLA™) by 3D Systems was the first commercially available layer-additive process for creating physical objects directly from CAD data. 3D Systems was founded in 1986 and their first SLA machine was built in 1987. At present, 3D Systems produces four different SLAs. The smallest is SLA 250 which utilises a helium-cadmium laser with wavelength of 325 nm. All the other SLAs from this company use solid state Nd: YVO_4 lasers emitting at 354.7 nm. Currently, the most popular system for RP and tooling is the SLA 3500 shown in Figure 3.1. The specifications of all 3D Systems SLA products are summarised in Figure 3.2.

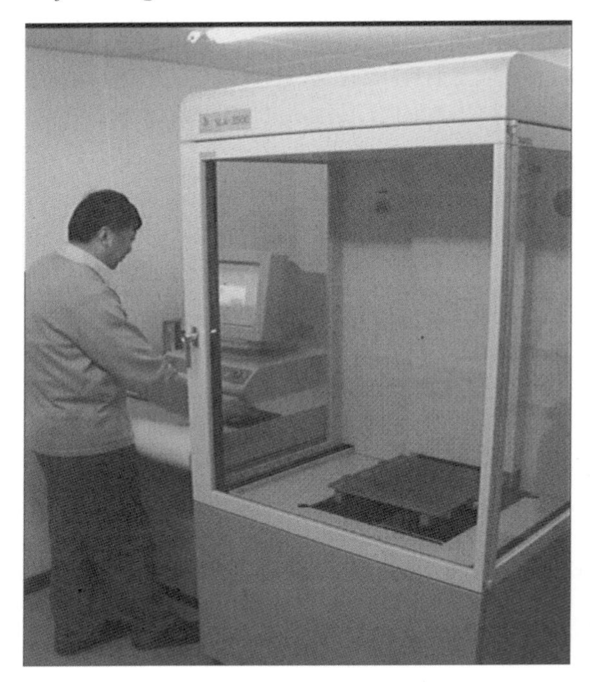

Figure 3.1 SLA 3500 RP system

A SLA creates solid or partially solid SL parts with either acrylic or epoxy resins in one of several build styles, the three most common being ACES™, STARWEAVE™ and QuickCast™ [3D Systems, 1996]. Completely hollow parts are not normally constructed as these are very fragile in the green state and deform with handling.

When using ACES™, the interior of the part is almost wholly cured by the laser (Figure 3.3). This is achieved by applying a hatch-spacing which is equivalent to half the line-width. This spacing is chosen so that all the solidified resin receives the same cumulative UV exposure and hence the downward facing surfaces are flat. This style may only be used with epoxy resins that do not shrink much when polymerised otherwise the connected lines would cause warping in the prototype. It is the most accurate of the three build styles for low-distortion resins and is employed when making high precision parts although the drawing time is the longest of the three styles [3D Systems, 1996; Jacobs, 1996].

	SLA 250/50HR	**SLA 3500**	**SLA 5000**	**SLA 7000**
Laser type, Wavelength, Power	HeCd, 325 nm, 6 mW	Solid state $Nd:YV0_4$, 354.7 nm, 160 mW	Solid state frequency tripled $Nd:YV0_4$, 354.7 nm, 216 mW	Solid state frequency tripled $Nd:YV0_4$, 354.7 nm, 800 mW
Layer Thickness*, mm	0.0625-0.1	0.05-0.1	0.05-0.1	0.0254-0.127
Beam diameter, mm	0.06-0.08	0.20-0.30	0.20-0.30	0.23-0.28 to 0.685-0.838
Drawing speed	635 mm/s	2.54 m/s	up to 5.0 m/s	2.54-9.52 m/s
Elevator Resolution and Repeatability, mm	0.0025	0.00177, ±0.005	0.00177, ±0.013	0.001, ±0.001
Max part weight, kg	9.1	56.8	68.04	68.04
Vat capacity, L	32.2	99.3	253.6	253.6
Max build envelope, mm	250 x 250 x 250	350 x 350 x 400	508 x 508 x 584	508 x 508 x 600
Operating system	MS-DOS	Windows NT	Windows NT	Windows NT
Size, m	1.24 x 0.69 x 1.64	0.95 x 1.02 x 2.00 & 0.85 x 1.02 x 1.03	1.88 x 1.19 x 2.02	1.88 x 1.22 x 2.03 & 1.12 x 1.22 x 1.02
Weight, kg	461	1100	1363	1455

* Geometry, build style, material and parameter dependent

Figure 3.2 Technical specifications of SLA systems [3D Systems, 2000]

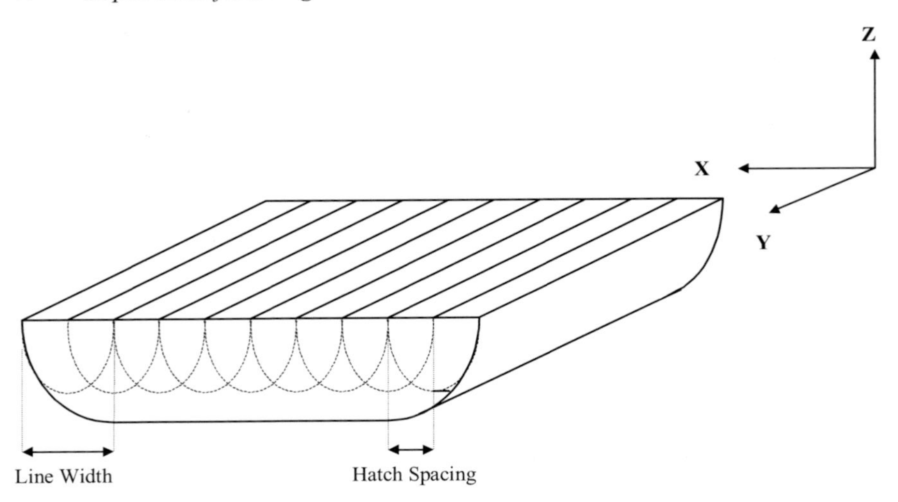

Figure 3.3 ACES™ build style: Repeated, even laser exposure produces a flat base

STARWEAVE™ provides stability to a solid part by hatching the interior with a series of grids which are offset by half of the hatch spacing every other layer (Figure 3.4). The grids are drawn such that the ends are not attached to the part border to reduce the overall distortion. Also to keep the distortion low, the grid-lines do not touch one another. However, they are located as close together as possible to improve the green strength of the part [3D Systems, 1996; Jacobs, 1996]. This build style should be employed with acrylic resins which shrink when polymerised. It is sometimes used with epoxy resins in preference to ACES™ because the draw time is lower.

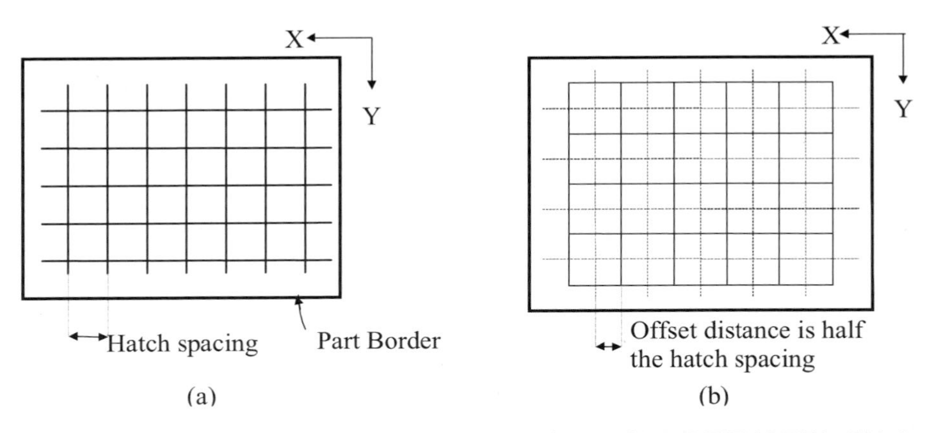

Figure 3.4 STARWEAVE™ build style: (a) One layer of STARWEAVE™. This is composed of a cross-hatched grid which is detached from the part border; (b) Alternate layers of STARWEAVE™ are offset by half the hatch-spacing

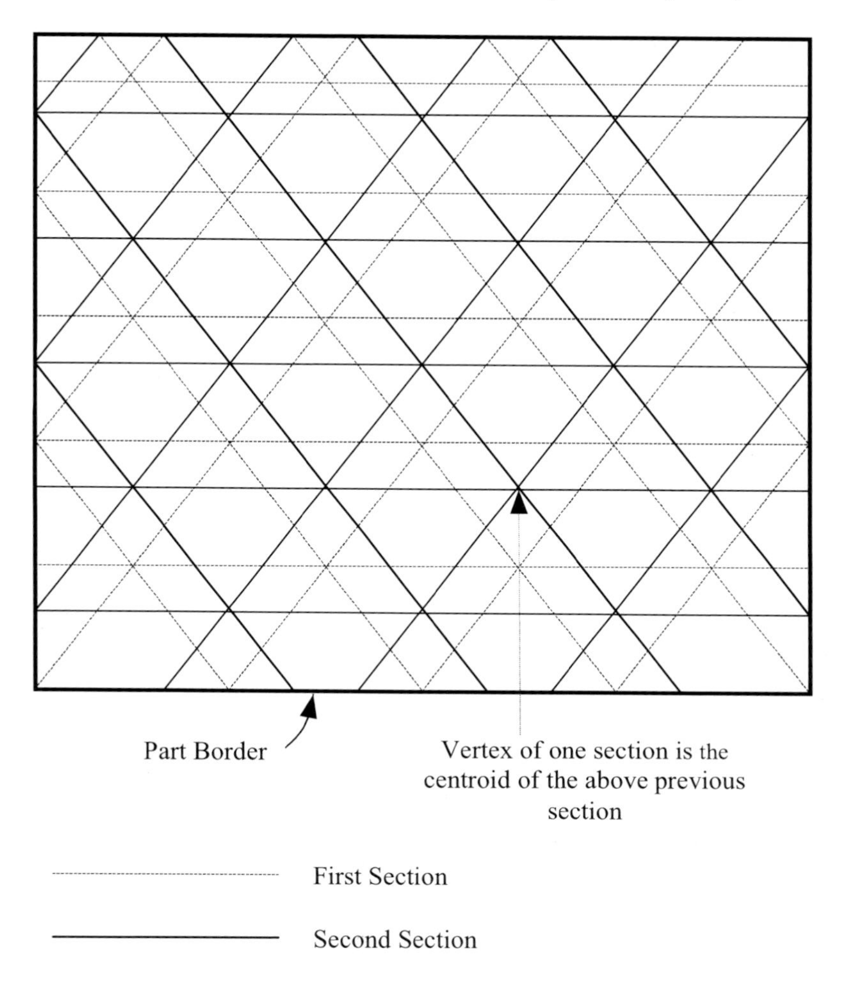

Figure 3.5 QuickCast™ build style: Parts are hatched with offset triangles

QuickCast™ is usually adopted when the prototype is to serve as a pattern for investment casting as it produces almost hollow parts. The outline of the layer is drawn before the interior is hatched. Either squares (QuickCast™ version 1.1) or equilateral triangles (QuickCast™ version 1.0) are used to fill the part and these are offset after a specified vertical build distance to facilitate resin drainage. The triangles are offset such that the vertices of one section are above the centroids of the triangles in the previous section (Figure 3.5). The squares are offset by half of the hatch spacing. Since squares have larger interior angles than triangles, the meniscus of resin will be smaller so better drainage is achieved [3D Systems, 1996; Jacobs, 1996]. Horizontal sections that form the outer surface of the part are completely solidified and are referred to as skinfill areas. Three layers are drawn with skinfill areas corresponding to the part surface to avoid the formation of 'pinholes' when the supports are removed and to prevent the upwards-facing horizontal surfaces from sagging [Jacobs, 1995; Jacobs, 1996]. These skinfills support the surface which means that the hatch spacing may be larger. It also means that a smaller percentage of the prototype is solid [Jacobs, 1996]. Vents and drains must be designed into these areas to allow the excess resin to bleed from the part. These parts will collapse quickly upon firing so that little stress is developed on the ceramic investment shell, thus preventing it from being damaged. Because QuickCast™ parts have a large surface area and the resin is hygroscopic, they should be used as quickly as possible or stored in an area with controlled humidity to prevent later distortion from water absorption.

A range of several resins has been developed by Ciba-Geigy in collaboration with 3D Systems especially for the SL process. The material properties of these resins are summarised in Figure 3.6.

Other companies have also developed RP systems based on the same principle of polymer curing by employing ultraviolet lasers. In particular, the following companies are producing SL systems: SONY, Computer Modelling and Engineering Technology Inc. (established by Mitsubishi Corp., NTT Data Corporation and Asahi Denka Kogyo K.K. in 1990), Teijin Seiki Ltd., Meiko Co Ltd. and AAROFLEX, Inc. [Aaroflex, 2000].

The advantages of SLA are that it yields a surface finish comparable to that of NC milling, is a well proven system with over 1000 machines in use worldwide and is reasonably fast and accurate [Dickens, 1995; Ippolito et al., 1995]. SL systems have an accuracy of ± 100 μm and can achieve layers 25 μm thick [3D Systems, 1996]. To utilize the resin vat fully and shorten production time, several parts may be built at once.

The disadvantages are that the material is expensive and toxic, has an unpleasant odour, and must be shielded from light to avoid premature polymerisation; there is also only a limited choice of resins. The parts may be brittle and translucent and they need supports which may adversely affect the surface finish when removed. Changing the resin in the vat is a lengthy and costly procedure.

		Resin Types									
	Measurement method	**5170[1]**	**5190[1]**	**5195**	**5210[2]**	**5220[3]**	**5430[3]**	**5510[4]**	**5520[5]**	**5530[3]**	**7510**
Density @ 25° C		1.14	1.15	1.16	1.15	1.14	1.181	1.13	1.15	1.188	1.17
Tensile strength	ASTM D 638, MPa	59-60	55-57	46.5	45	62	51-56	77	26-33	56-61	57
Tensile modulus	ASTM D 638, MPa	3737-4158	2172-2275	2090	3020	2703	2358-2668	3296	1034-1379	2889-3144	2634
Elongation at break	ASTM D 638, %	7-19	9	11-22	1.6	8.3	2.9-4.9	5.4	23-43	3.8-4.4	10.1
Flexural strength	ASTM D 790, MPa	107-108	75-90	49.3	74	94	109-113	99	29-41	63-87	81
Flexural modulus	ASTM D 790, MPa	2920-3006	2110-2450	1628	3061	2951	3089-3165	3054	689-896	2620-3240	2386
Impact strength, notched	ASTM D 256, J/m	27-30	27	54	27	37	21	27	59	21	37.40
Hardness	DIN 53505, Shore D	85	80	83	85	86	87	86	80	88	87
Glass deflection temp	DMA, E" peak (UV postcure), ASTM D 648				92°C	53°C	88° C	68° C	34° C		
Heat deflection temp	ASTM D 648 @ 0.455 MPa	55° C	50°C	47° C	99°C		63° C	50° C	44° C	80° C	58° C
	ASTM D 648 @ 1.822 MPa	49° C	43°C		74°C	42°C	52° C	43° C			49° C

1. 60 minutes UV postcure (ACES build style)
2. 90 minute UV postcure + 2 hrs @ 120°C thermal postcure
3. 90 minute UV postcure only
4. 90 minute UV postcure + 2hrs @ 80° C thermal postcure
5. 1.5-6 hours UV postcure; higher values are obtained with longer UV postcure times

Figure 3.6 Properties of SL resins [3D Systems, 2000]

3.2 Solid Ground Curing Systems (Cubital Ltd.)

SGC systems were developed by Cubital Ltd. Cubital was founded in 1987 as an internal R&D unit at Scitex Corporation Ltd. and the first SGC system was installed in 1991. The company produces two types of systems based on the SGC model-making technology, Solider 4600 – an entry level RP machine, and Solider 5600 for users with high-capacity requirements. The Cubital's Solider 4600 system is shown in figure 3.7. Figure 3.8 summarises the technical characteristics of both systems.

An advantage of SGC systems is that the entire layer is solidified at once, reducing the part creation time. All the resin within a layer is completely cured by this method, so parts may be more durable than the hatched prototypes created using other processes and no postcuring is required. Supports are unnecessary because wax is used to fill the gaps in the resin and also to brace the part [Cubital, 2000].

Figure 3.7 Cubital's Solider 4600 system (Courtesy of Cubital Ltd)

The disadvantages of SGC are that the machine is noisy and large and needs to be constantly manned. It wastes a large amount of wax since the used wax cannot be recycled after being removed in a hot water or citric acid bath. SGC systems are also prone to breakdowns [Jacobs, 1996; Waternam and Dickens, 1994; Kai, 1994]. The resin models produced using SGC are solid and so are not suitable for later investment casting. This is because the coefficient of thermal expansion of the resin is an order of magnitude greater than that of the ceramic system so the ceramic moulds will crack when the sacrificial part is burnt out [Waterman and Dickens, 1994].

	SGC 4600	**SGC 5600**
Working Volume	350x350x350 mm	500x350x500 mm
Surface Definition	0.15 mm	0.15 mm
X-Y Resolution	0.084 mm	0.084 mm
Layer Thickness	0.15 mm	0.1 - 0.2 mm
Min Feature Size	0.4 mm	0.4 mm
Net Production Rate	164 cm^3/hour	426 cm^3/hour
Layer Build Speed	120 s/layer	70 s/layer
Control system	Unix based workstation	Unix based workstation
Price	$275,000	$470,000

Figure 3.8 Specifications of SGC systems [Cubital, 2000]

The resolution (minimum feature size) of this system is 400 μm in the X-Y plane and 100 μm in the Z direction. The least expensive SGC machine costs around $275,000 and weighs about 5000 kg. The largest build chamber available is 500 x 350 x 500 mm. Typically, a layer can be built in 70-120 s, depending on the machine used. Of this building time, 3 s are for exposing the layer to a 2000W UV lamp, the remaining time being needed to clear the part of resin and to add, chill and mill the wax [Cubital, 2000].

3.3 Fused Deposition Modelling Systems (Stratasys, Inc.)

The FDM process was developed in 1988 by S. Crump who founded Stratasys the following year. Since then the company has produced more than 1000 systems. The first product, the 3D MODELER, was introduced by the company in 1992. Stratasys's FDM current product range includes the following RP systems: FDM 2000, FDM 3000, FDM 8000 and FDM Quantum. All four FDM systems include two nozzles, one for the part material and one for the support material. The latter is cheaper and breaks away from the prototype without impairing its surface [Statasys, 2000]. It is also possible to create horizontal supports to minimise material usage and build time [Crump, 1991]. The technical characteristics of these systems are summarised in Figure 3.9. The most popular system, the FDM 2000 (Figure 3.10), builds functional prototypes or casting masters up to 254x254x254 mm in size. The system costs around $87,000 and has an accuracy of ± 127 μm [Stratasys, 2000].

	FDM 2000	**FDM 3000**	**FDM 8000**	**Quantum**
Build Size, mm	254 x 254 x 254	254 x 254 x 406	457 x 457 x 609	600 x 500 x 600
Accuracy	± 0.127 mm	± 0.127 mm	± 0.127 - 0.254 mm	± 0.127mm (up to 127 mm) and ± .0015 mm/mm (greater than 127 mm)
Size, mm	W660 x H914 x D1067	W660 x H1067 x D914	W1486 x H1905 x D1003	W2235 x H1981 x D1118
Weight	160kg	160kg	392 kg	1134 kg
Power Require-ments	220-240 VAC, 50/60 Hz, 10A single phase	208-240 VAC, 50/60 Hz, 10A single phase	220-240 VAC, 50/60 Hz, 10A single phase	208-240 VAC, 50/60 Hz, 50A single phase
Materials	ABS (White) ABSi Investment Casting Wax Elastomer	ABS (White) ABSi Investment Casting Wax Elastomer	ABS	ABS
Layer Width	0.254 to 2.54 mm	0.254 to 2.54 mm	0.254 to 2.54 mm	0.38 to 0.51 mm
Layer Thickness	0.05 to 0.762 mm	0.05 to 0.762 mm	0.05 to 0.762 mm	0.18 to 0.25 mm

Figures 3.9 Technical characteristics of FDM systems [Stratasys, 2000]

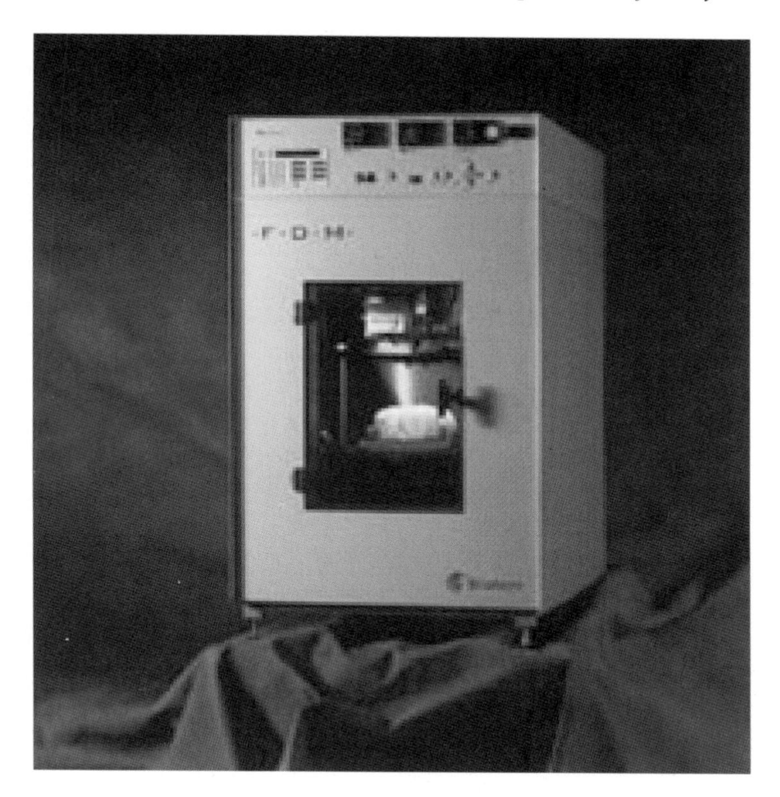

Figure 3.10 FDM 2000 system (Courtesy of Stratasys, Inc.)

In January 1998, Stratasys introduced the FDM Quantum system which incorporates the MagnaDrive technology. This technology uses an X-Y electro-magnetic motion-control system in combination with dual-axis linear motors. The combination of these technologies provides precise and repeatable two-axis motion control in a single plane without requiring a gantry [Stratasys, 2000]. This new design eliminates the typical moving parts of a gantry system, such as cables, belts and pulleys thereby simplifying the mechanics and increasing reliability. The MagnaDrive technology allows coordinated moves such as contouring or circular interpolation, to be realised with high precision. The MagnaDrive heads and FDM Quantum system are shown in Figures 3.11 and 3.12.

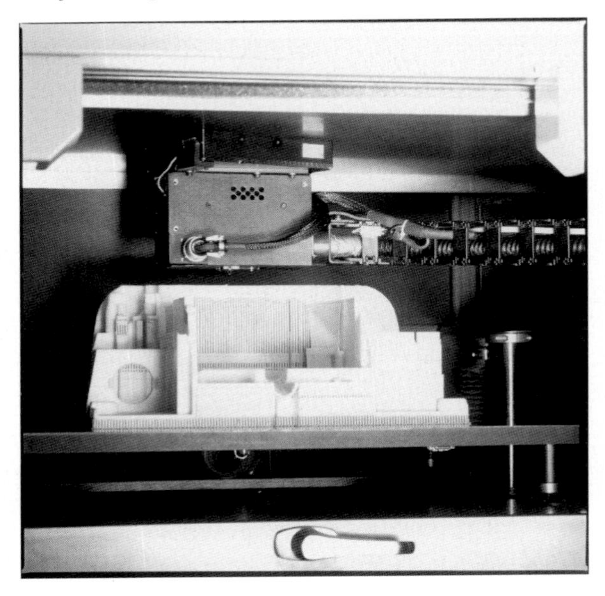

Figure 3.11 FDM Quantum extrusion heads employing MagnaDrive technology (Courtesy of Stratasys, Inc.)

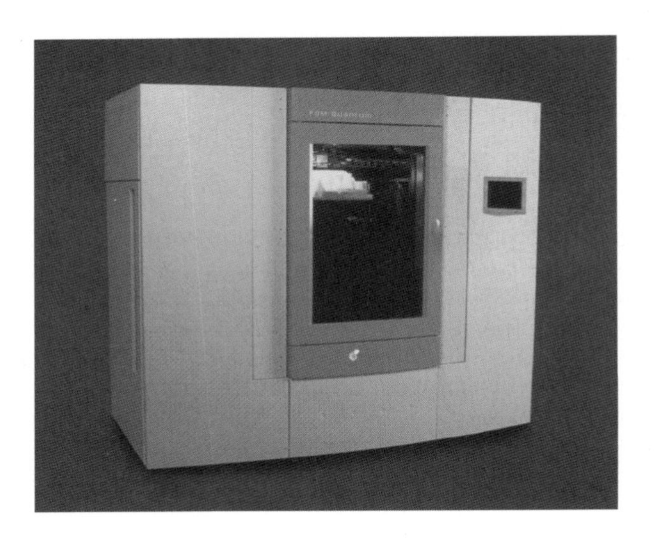

Figure 3.12 FDM Quantum system (Courtesy of Stratasys, Inc.)

In August 1999, Stratasys introduced the FDM 3000 system which incorporates a new support removal method for use with an ABS modelling material called WaterWorks. With WaterWorks, a completed model with supports is immersed in a water-based solution which dissolves the support material and leaves a clean model with smooth surfaces [Stratasys, 2000]. The FDM 3000 system costs in the range $125,000-$135,000 depending on options.

There are a number of thermoplastic modelling materials for FDM systems. Currently, the materials available commercially are ABS, a high impact grade of ABS, investment casting wax and an elastomer [Stratasys, 2000]. The FDM process utilises a spool-based filament system and therefore all materials are provided on spools.

An advantage of a FDM system is that it may be viewed as a desktop prototyping facility in a design office since the materials used are cheap, non-toxic, not smelly and environmentally safe. A range of materials is available as already mentioned. Parts made by this method have a high stability since they are not hygroscopic [Stratasys, 2000].

The disadvantage is the surface finish of the parts, which is inferior to that produced using SL. This is because the resolution of the process is lower as this is dictated by the filament thickness [Bidanda, 1994]. It has not yet been demonstrated whether the material extrusion may be stopped quickly enough to produce small holes in vertical sections [Jacobs, 1996].

3.4 Selective Laser Sintering Systems (DTM Corp. and EOS GmbH)

The SLS process was developed and originally patented by the University of Texas at Austin. To commercialise the SLS process, the DTM Corporation was formed in Austin, Texas. DTM now owns 79 patents worldwide covering different aspects of the SLS process, systems and materials. The first commercially available system was introduced by DTM in 1992. The latest SLS system by DTM is the Sinterstation 2500 Plus (Figure 3.13) with build chamber dimensions of 381 mm in width, 330 mm in depth and 457 mm in height. A summary of the Sinterstation 2500 Plus specifications is given in Figure 3.14. The Sinterstation 2500 Plus costs around $300,000.

A large range of materials is available for this process and basically any substance that can be pulverised to a fine powder may be employed. At present, nylon (polyamides), nylon composites, polystyrene and polycarbonate are in use [DTM, 2000a]. These are cheaper than the resins for SL, non-toxic, and safe and may be sintered with relatively low-power lasers (10 - 20 W). Metal powders and sand, coated with a suitable binder layer, can also be sintered for direct production of metal tooling inserts and sand cores and cavities. More information about the rapid tooling applications of the SLS process can be found in Chapter 7.

Figure 3.13 Sinterstation 2500 Plus (Courtesy DTM Corp.)

Currently, two nylon-based materials are available commercially for the SLS process (DuraForm Polyamide and DuraForm Glass-filled). Typically, the nylon-based materials are used for the production of prototypes for testing or parts with working features such as hinges, snap fits or clips. DuraForm prototypes can be relatively easily finished to a smooth appearance [DTM, 2000b]. The production of Nylon parts is generally cost effective when a small number (1-5) of parts is required. The properties of nylon materials are summarised in Figure 3.15. An example of a part built in DuraForm PA is shown in Figure 3.16.

Another group of SLS materials is used for producing casting patterns. Currently, two materials, TrueForm and CastForm, are employed for building patterns. TrueForm, an acrylic-based powder, is processed at relatively low temperatures compared with nylon which limits shrinkage to 0.6% [Van de Crommert et al., 1997] and is for making parts with good accuracy but moderate strength. The density of TrueForm parts can vary from 70 to 90% depending on build parameters and they can be polished to a mirror-like finish. CastForm, a polystyrene-based powder, was introduced by DTM in 1999. This new material offers significant advantages over TrueForm when employed to produce patterns for investment casting. In particular, it features a low ash content and is compatible with standard foundry practices. Processing CastForm creates porous low-density patterns that are subsequently infiltrated with a low-ash foundry wax to yield patterns containing 45% polystyrene and 55% wax. The material properties of TrueForm and CastForm are summarised in Figure 3.17.

Build Chamber Dimensions	381^W x 330^D x 457^H mm
Laser	50- or 100-Watt CO_2, (Beam Diameter = 0.420 mm)
Beam Delivery System	M3ST Galvanometers, 3 axis Scan Speed = 5,000mm/sec Positional Accuracy = 50mm
Software	Windows NT OS DTM Sinterstation System Software Materialize NV, Magics RP® Software
Computer System	Pentium-based controller
System Dimensions	Process Station = 2133^W x 1346^D x 1981^H mm Computer Cabinet = 609^W x 609^D x 1828^H mm Chiller = 533^W x 838^D x 914^H mm
Peripherals	Breakout Station / Air Handler / Sifter, Vacuum Cleaner
Power Requirements	240 VAC, 12.5 KVA, 50/60 Hz, 3-phase
Nitrogen Requirements	Purity = 99.9% Minimum Pressure = 1.7 bars Continuous Flow Rate = 19 lpm
Atmospheric Requirements	Temperature Range = 15° to 27° C Relative Humidity = <60%

Figure 3.14 Sinterstation 2500^{plus} system specifications

SLS materials are cheaper than the resins used for SL, non-toxic and safe and may be sintered with relatively low-powered lasers. However, nylon parts need a long cooling cycle in the machine before they can be removed. For example, nylon composite parts require 6-8 hours to cool down. The materials employed by the system are sensitive to the different heating and laser parameters and each material requires distinct settings. These can be difficult and time-consuming to set.

General Properties	Unit	ASTM Test Method	DuraForm Polyamide	DuraForm Glass-Filled
Specific Gravity, 20°C		D792	0.97	1.40
Moisture Absorption, 23°C	%	D570	0.41	0.30
Powder Density, Tap	(g/cm3)	D4164	0.59	0.84
Average Particle Size	µm	Laser Diffraction	58	48
Particle Size Range, 90%	µm	Laser Diffraction	25-92	10-96
Thermal Properties				
Melting Point, Tm	°C	DSC	184	185
DTUL, 0.45 Mpa	°C	D648	177	175
DTUL, 1.82 Mpa	°C	D648	86	110
Mechanical Properties				
Tensile Strength at Yield	MPa	D628	44	38.1
Tensile Modulus	MPa	D628	1,600	5,910
Tensile Elongation at Break	%	D628	9	2
Flexural Modulus	MPa	D790	1,285	3,300
Impact Strength				
Notched Izod	J/m	D256	214	96
Unnotched Izod	J/m	D256	428	101

Figure 3.15 Properties of DuraForm materials

Figure 3.16 Boeing air vehicle component (Courtesy of DTM Corp)

A drawback of the SLS process is that the recycled powders require sieving to ensure that no globules are present that would interfere with the smooth application of the next powder layer. The system also requires an inert Nitrogen atmosphere in which to sinter the materials.

	Units	Test Method	TrueForm	CastForm
Powder Properties				
Average Particle Size	μm	Laser Diffraction	33	62
Particle Size Range, 90%	μm	Laser Diffraction	15-60	25-106
Powder Density, Tap	g/cm3	ASTM D4164	0.3-0.4	0.46
Thermal Properties				
Glass Transition: Tg	°C	DSC	69	89
DTUL 0.45 Mpa	°C	ASTM D648	62	33
DTUL, 1.82 Mpa	°C	ASTM D648	59	40
Physical and Mechanical Properties				
Specific Gravity, 20° C		ASTM D792	1.08	0.86
Moisture Absorption, 20°C, 65% R.H.	%	ASTM D570	0.26	0.06
Ash Content	%	ASTM D482	1.1	0.02
Tensile Strength	MPa	ASTM D638	10.0	2840
Tensile Modulus	MPa	ASTM D638	1.1	1604
Tensile Elongation at Break	%	ASTM D638	1.2	
Impact Strength				
Notched Izod	J/m	ASTM D256	8.1	11
Unnotched Izod	J/m	ASTM D256	21.9	14
Surface Finish				
As Processed (Ra)	μm	DTM	5.5	13
After Polishing (Ra)	μm	DTM	0.7	3

Figure 3.17 Material properties of TrueForm and CastForm

Laser sintering systems are also manufactured by EOS GmbH Electro Optical Systems. The company introduced EOSINT P machines in 1994. The machines are designed for fabricating physical prototypes from thermoplastic powder materials such as polystyrene and nylon. The principal applications of EOSINT P are for building investment casting patterns and functional test-parts. A summary of specifications of EOSINT P350 is given in Figure 3.18.

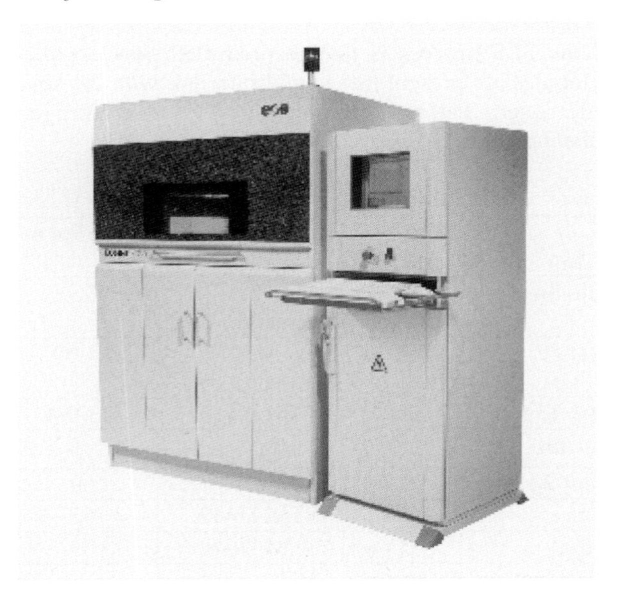

Building volume	340 x 340 x 600 mm³
Laser type	CO_2, 50 W
Laser scan speed	up to 5 m/s
Building speed	10-25 mm height/h (material dependent)
Layer thickness (typ)	0.1-0.2 mm
Process computer	PC
Electrical supply	400 V, 32 A
Compressed air supply	minimum 5.000 hPa; 6 m³/h
Cooling water supply	1.500 hPa - 4.000 hPa; 5 l/min
Dimensions	Process cabinet 1.250 x 1.300 x 2.150 mm Control cabinet 610 x 820 x 1.785 mm
Weight	approx. 800 kg
Workstation	Silicon Graphics Indigo
PC	Windows 95, Windows NT
Interface to CAD	Standard: STL, CLI Optional: VDA-FS, IGES, CATIA

Figure 3.18 Specification of EOSINT P350 (Courtesy of EOS GmbH)

3.5 Laminated Object Manufacturing Systems (Helisys, Inc.)

The LOM process was developed by Helisys Inc. The company shipped the first commercial LOM system in 1991. Currently, Helisys produces two models of the LOM system, LOM-1015 Plus and LOM-2030H (Figure 3.19) [Helisys, 2000]. These two systems employ a 25 and 50 W CO_2 laser respectively to cut the material. A summary of their specifications is given in Figure 3.20.

Figure 3.19 LOM-2030H system

No correction is applied to the width of the CO_2 laser beam as yet, which means that parts will have a small error associated with their dimensions. Smaller hatches must be used on up- and down-facing surfaces to facilitate the removal of the waste material that has bonded to the part. The sequence of separating a LOM object from the surrounding excess material is shown in Figure 3.21. It may also be necessary to stop the build to excavate paper from otherwise hard-to-access places. Once the parts have been completed, they have to be sealed with a urethane, silicon or epoxy spray to prevent later distortion of the prototype due to water absorption. The height is measured and the cross-sections are calculated in real time to correct any errors in the build direction [Jacobs, 1996].

	LOM-1015 Plus	LOM-2030H
Part envelope, mm	381^{L} x 254^{W} x 356^{H}	813^{L} x 559^{W} x 508^{H}
Laser power, W	25	50
Laser beam diameter, mm	0.20-0.26	0.203-0.254
Motion system repeatability, mm	0.05	0.0508
System controller	Pentium-based	Pentium-based
System software	Windows NT OS LOMSlice™	Windows NT OS LOMSlice™
Material thickness, mm	0.08-0.25	0.076-0.254
Material size, mm	356 mm roll width, 356 mm roll diameter	711 mm roll width, 711 mm roll diameter
Dimensions, cm	122.6^{L} x 74.3^{W} x 130.8^{H}	206^{L} x 141^{W} x 140^{H} and 59^{L} x 83^{W} x 140^{H}
Weight, kg	454	1,288 and 141
Power requirements	220 VAC, 15 A, 50/60 Hz, single phase	220 VAC, 30 A, 50/60 Hz, single phase
Atmospheric Requirements	Temperature = 20° to 27° C Humidity = <50%	Temperature = 20° to 27° C Humidity = <50%

Figure 3.20 A summary of LOM system specifications [Helisys, 2000]

The minimum layer thickness that LOM machines can handle is 25 - 127 μm and their maximum accuracy is ±127 μm. The maximum cutting speed achievable is 380 mm per second [Waterman and Dickens, 1994; Anon 1993]. An example of a LOM part is shown in Figure 3.22. A wide range of relatively cheap materials is available – for instance, parts may be made using paper.

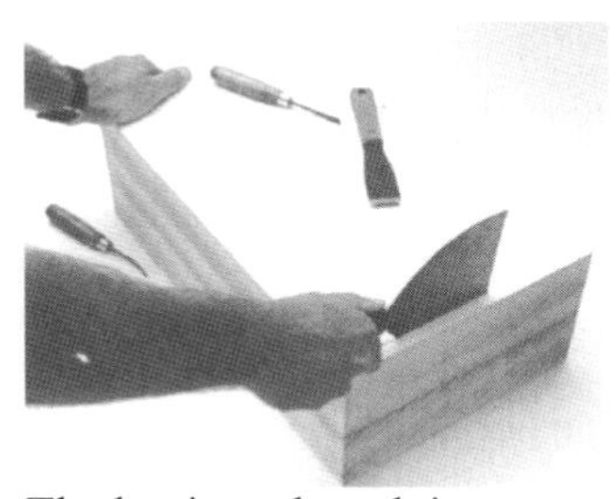

The laminated stack is removed from the machine's elevator plate

The surrounding wall is lifted off the object to expose cubes of excess material

Cubes are separated from the object's surface

The object's surface can then be sanded, polished or painted as desired

Figure 3.21 Separation of a LOM part (Courtesy of Helisys, Inc.)

Figure 3.22 Real Engine Power Take-Off housing (Courtesy of Helisys, Inc.)

3.6 Paper Lamination Technology (Kira Corp)

The PLT process was developed by KIRA Corporation. The company has been producing RP machines since 1994. The main differences between the LOM and PLT processes are in the material used and the methods employed for cutting the contours of the part cross-sections, which are a CO_2 laser and computerised knife respectively. Two versions of the PLT process have been implemented. The first system, PLT-A4 (Figure 3.23), selectively prints a resin toner on plain paper sheets and then hot presses the sheets to laminate them together. The second system, PLT-A3, employs a roll of paper. A summary of their specifications is given in Figure 3.24.

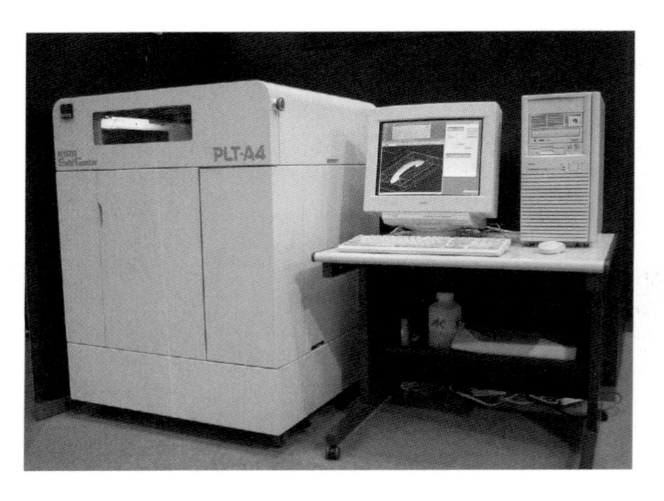

Figure 3.23 PLT-A4 system (Courtesy of Kira Corp)

System	**PLT-A4**	**PLT-A3**
Material	Sheet paper	Roll paper
Max. build size (mm)	190x280x200	400x280x300
Paper thickness (mm)	0.08 or 0.15	0.08 or 0.15
Resolution (mm)	XY ± 0.05, Z ± 0.1	XY ± 0.05, Z ± 0.1
Accuracy (mm)	± 0.2	± 0.2
Power supply	AC100V±10V (50/60 Hz ± 1Hz)	3 Phase AC200V±10V (50/60 Hz ± 1Hz)
Weight (kg)	450	550

Figure 3.24 Specifications of PLT systems [KIRA, 2000]

3.7 Laser Engineering Net Shaping (LENS^{TM}) Systems (Optomec Design Co.)

The LENS^{TM} process was initially developed at Sandia National Laboratories. The process was commercialised by Optomec Design Co in 1997 through a licensing agreement. The process builds parts in an additive manner from powdered metals using a high-powered Nd:YAG laser. The laser beam is focused onto a substrate where metallic powder is injected under computer guidance to build up three-dimensional parts (Figure 3.25). The parts are fabricated vertically, one layer at a time. The visible glow in the active area is from the heated metal. Due to the rapid solidification that occurs during the process, superior strength and ductility are achieved for most metal alloys. No further heat treating or cooling is required. The process has been demonstrated to yield a dimensional accuracy of ±127 μm and a surface finish after some post processing of R_a 0.25 [Optomec, 2000].

Two systems are commercially available, LENS 750 and 850 (Figure 3.26). A summary of their technical specifications is given in Figure 3.27. LENS technology has been applied to a broad range of metals and alloys including: 304 and 316 Stainless Steel; Iron-Nickel Alloys; H13 and MM10 Tool Steels; 625, 690 and 718 Inconel; Titanium Alloys; Tungsten; Haynes 230; Nickel Aluminide. The LENS processed materials can have a significantly greater strength and ductility compared to similar composition materials produced by conventional processes. Another advantage of this process is its ability to use non-weldable materials like MAR-M 247, RHENIUM and many Nickel based superalloys.

Figure 3.25* LENS^{TM} process (Courtesy of Optomec Design Co)

The LENS process has been applied successfully to tooling, repair and rebuilding, functional prototyping and short run manufacturing (Figure 3.28). The process is under continuous development to widen its area of applications.

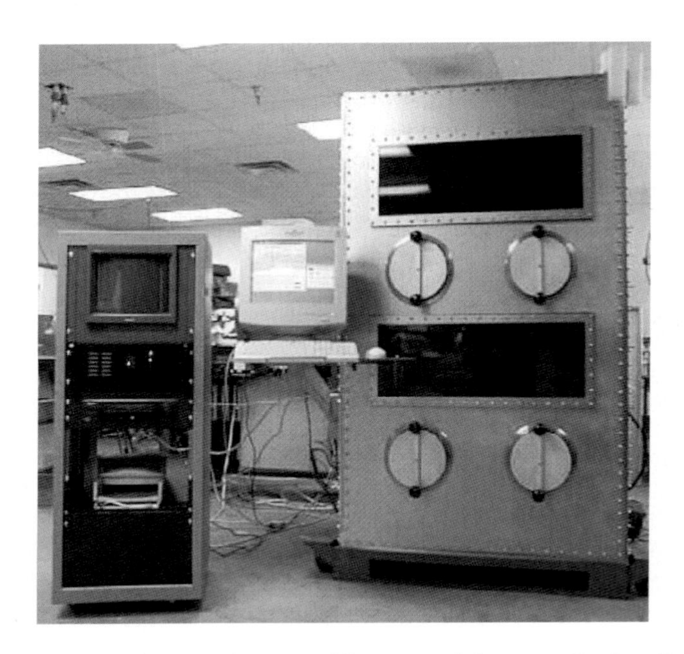

Figure 3.26 LENS 850 system (Courtesy of Optomec Design Co)

Model	**LENS 750**	**LENS 850**
Laser system	750 W cw Nd:YAG, upgradable to 1400 W	1100 W dual head cw Nd:YAG, upgradable to 2200 W
Build Chamber	12" x 12" X-Y motorised stages, 12" Z axis, Controlled atmosphere (argon re-circulation system)	18" x 18" X-Y gantry stage, 42" Z axis, Controlled atmosphere (argon re-circulation system)
Powder delivery	Pneumatic powder delivery system	Dual pneumatic powder delivery systems
Power requirements	208V/3 Phase/100A	208V/3 Phase/100A
System control	Pentium-based system	Pentium-based system

Figure 3.27 Technical specifications of LENS systems [Optomec, 2000]

Figure 3.28* Parts produced employing LENS and stereolithography (Courtesy of Optomec Design Co)

3.8 Summary

This chapter has described the technical characteristics of commercially available RP systems. In addition, their technological capabilities have been discussed and their possible application areas outlined.

References

AAROFLEX Web page (2000) **AAROFLEX, Inc**., 8550 Lee Highway, Suite 650, Fairfax, Virginia 22031 USA, http://www.aaroflex.com/.

Anon (1993) State of the Art Review-93-0, **MTIAC**, 10 West 35 Street, Chicago, IL 60616, USA.

Bidanda B, Narayanan V and Billo R (1994) Reverse Engineering and Rapid Prototyping, **Handbook of Design, Manufacture and Automation**, eds. Dorf RC and Kusiak A, Wiley, NY, USA, pp 977–991.

Crump SS (1991) Fast, Precise, Safe Prototypes with FDM, **ASME Annual Winter Conference**, Atlanta, December, Vol. 50, pp 53–60.

Cubital Web page (2000) **Cubital Ltd**., 13 Ha'Sadna St., Industrial Zone North, Ra'anana, 43650, Israel, http://www.cubital.com/.

Dickens PM (1995) Research Developments in Rapid Prototyping, **Proc. IMechE, Part B: Journal of Engineering Manufacture**, Part B, Vol. 209, pp 261-266.

DTM Web page (2000a) **DTM Corporation**, 1611 Headway Circle, Building 2, Austin, Texas 78754, USA, http://www.dtm-corp.com.

DTM White paper (2000b) on functional prototyping with DuraForm and SLS, **DTM Corporation**, 1611 Headway Circle, Building 2, Austin, Texas 78754, USA, http://www.dtm-corp.com/products/dura2.html.

Ippolito R, Iuliano L and Gatto A (1995) Benchmarking of Rapid Prototyping Technologies in Terms of Dimensional Accuracy and Surface Finish, **CIRP Annals**, Vol. 44, 1, pp 157-160.

Jacobs PF (1995) QuickCast™ 1.1 and Rapid Tooling, **4th European Conference on Rapid Prototyping and Manufacturing**, Nottingham, June 13-15, pp 1-27.

Jacobs PF (1996) Stereolithography and other RP&M technologies, **Society of Manufacturing Engineers - American Society of Mechanical Engineer**.

Helisys Web page (2000) **Helisys, Inc**., 24015 Garnier Street, Torrance, California 90505-5319, USA, http://helisys.com/.

Kai CC (1994) 3D Rapid Prototyping Technologies and Key Development Areas, **Computing and Control Engineering Journal**, August, pp 200–206.

KIRA Web page (2000) **KIRA Corporation**, Omiyoshishinden, Kira-Cho, Hazu-Gun, Aichi, Japan, www.kiracorp.co.jp.

MIT Web page (1999) **MIT**, Three Dimensional Printing Group, http://me.mit.edu/groups/tdp/.

Optomec Web page (2000) **Optomec Design Company**, 2701-D Pan American Freeway - Albuquerque, New Mexico - 87107, USA, http://www.optomec.com/

Pham DT and Gault RS (1998) A Comparison of Rapid Prototyping Technologies, **International Journal of Machine Tools and Manufacture**, Vol. 38, pp 1257-1287.

Stratasys Web page (2000) **Stratasys, Inc**. 14950 Martin Drive, Eden Prairie, MN 55344-2020 USA, www.stratasys.com.

Van de Crommert S, Seitz S, Esser KK and McAlea K (1997) Sand, Die and Investment Cast Parts via the SLS Selective Laser Sintering Process, **DTM press release**, DTM GmbH, Hilden, Germany .

Waterman N.A. and Dickens P (1994) Rapid Product Development in the USA, Europe and Japan, **World Class Design To Manufacture**, Vol. 1, 3, pp 27–36.

3D Systems (1996) **3D Systems**, Worldwide Corporation HQ, 26081 Avenue Hall, Valencia, California, Maestro Workstation User Guide.

3D Systems (1996) 3D Systems, Worldwide Corporation HQ, 26081 Avenue Hall, Valencia, California, **The Edge**, Vol. V, 1.

Chapter 4 Technical Characteristics and Technological Capabilities of Concept Modellers

This chapter gives a technical overview of commercially available Concept Modellers. These are a new range of RP systems addressing the specific needs of CAD offices. Although CAD systems have empowered designers with a number of tools to minimise errors and maximise design quality offering facilities such as photorealistic visualisation, interactive product simulation, assembly analysis, and kinematic and stress analysis, the design remains intangible until a physical model is built. Concept Modellers fill this gap by offering relatively quick and cost effective methods for building physical models at any design stage. They are marketed as new CAD peripheral solutions which enable designers to verify and iterate their designs without leaving the office. Typically, Concept Modellers build models more quickly but not so accurately as other RP systems and usually cost less than $50,000. The following RP systems are classed as Concept Modellers:

- 3D Systems ThermoJet printer (Multi-Jet Modeller);
- Sanders ModelMaker II (Inkjet Modelling Technology);
- Z-Corporation Z402 (3D printer);
- Stratasys Genisys Xs printer;
- JP System 5;
- Objet Quadra system.

This chapter examines the technical characteristics, materials and application areas of Concept Modellers and reviews their main strengths and weaknesses. The chapter also gives examples of different RP applications in which these commercially available systems have been successfully employed.

4.1 3D Systems ThermoJet™ printer

The Multi-Jet Modelling (MJM) process was developed by 3D Systems in 1995. This technology complements 3D Systems's established line of Stereolithography products. Initially, the MJM machine was marketed as the Actua 2100 but since 1998 it has been known as the ThermoJet printer. The ThermoJet printer is shown in Figure 4.1 and the system specifications in Figure 4.2.

Figure 4.1 3D Systems ThermoJet printer (Courtesy of 3D Systems)

MJM parts are constructed from a thermoplastic material. The parts have a layer thickness of 40 μm, an X-Y resolution of 85 μm and a droplet placement accuracy of ± 100 μm [3D Systems, 1996].

Resolution	300 DPI
Maximum Model Size	250 x 190 x 200 mm (10 x 7.5 x 8 in)
Modeling Material	ThermoJet™ 88 and TJ2000 thermoplastic
Material Color Options	Neutral, grey, or black
Material Capacity	5.9 kg
Material Loading	2.3 kg cartridge
Interface	Ethernet 10/100 Base-TX, RJ-45 Cable, TCP/IP protocol
Platform Support	Silicon Graphics IRIX v6.5.2 Hewlett Packard HP-UX v10.2 ACE Sun Microsystems Solaris v2.6.0 IBM RS 6000 AIX v4.3.2 Windows NT v4.0
Power Consumption	230 VAC, 50/60 Hz, 6.3 Amps
Dimensions	1370 x 760 x 1120 mm
Weight	375 kg

Figure 4.2 Specifications of the ThermoJet printer [3D Systems, 2000]

As a result of the MJM raster-based building style, complex models can be built as quickly as simple models of similar overall dimensions. The model envelope is the determining factor for part build time [3D Systems, 2000]. Another advantage of the ThermoJet printer is that it is equipped with a cartridge system for easy loading of the thermopolymer material. Examples of models built using the printer are shown in Figure 4.3.

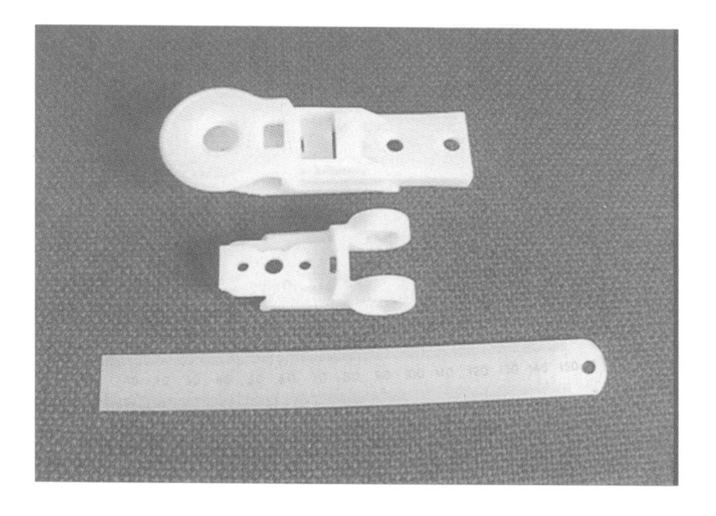

Figure 4.3* Example of model built using the ThermoJet printer

4.2 Sanders ModelMaker II (Inkjet Modelling Technology)

The inkjet modelling process was developed by Sanders Prototype Inc (SPI) in 1994. This technology combines a proprietary thermoplastic ink jetting technology with high-precision milling to build models or patterns that have a dimensional accuracy of ± 13 μm over 229 mm in the Z axis and up to ± 0.025 mm over 76 mm in the X-Y plane [Sanders, 2000]. The latest system developed by SPI is called PatternMaster. In terms of achievable accuracy, this system is superior to other Concept Modellers. The PatternMaster system is shown in Figures 4.4 and 4.5, and its specifications in Figure 4.6.

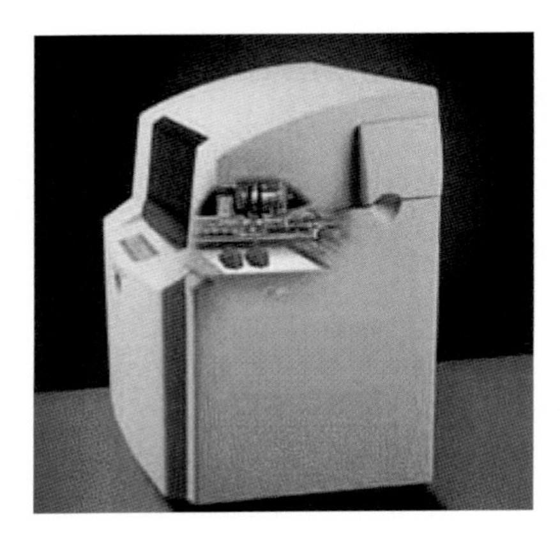

Figure 4.4 PatternMaster system (Courtesy of Sanders Prototype Inc.)

Figure 4.5 PatternMaster system detail (Courtesy of Sanders Prototype Inc.)

The PatternMaster systems use a proprietary construction thermoplastic material called ProtoBuild™. This material is compatible with investment casting and other mould making processes [Sanders, 2000]. For building support structures, the systems also employ SPI's proprietary wax material, ProtoSupport™, designed to be removed with a solvent without damage to the model.

This RP technology is ideal for building small models with intricate features, for example, in jewellery manufacture (Figure 4.7). Another important application area for this technology is the fabrication of precision investment cast components for aerospace, electronics and medical applications.

Build envelope	304.8 x 152.4 x 228.6 mm
Layer thickness	0.013 mm - 0.13 mm
Minimum feature size	0.25 mm
Dimensional accuracy	± 0.025 mm per 25 mm in X, Y and Z axes
Surface finish	32-63 micro-inches (RMS)
Wall thickness	1 mm
Size of micro-droplet	0.076 mm
Plotter carriage speed	Up to 500 mm per second
Footprint	889 (Width) x 660 (Depth) mm
Supported formats	.STL, .SLC, .DXF, .OBJ, .HPP (HPGL)
System controller	IBM-compatible PC running Microsoft Windows 95 or NT
Power requirements	115V 60Hz or 230V 50Hz AC

Figure 4.6 System specifications of PatternMaster [Sanders, 2000]

Figure 4.7* A ring produced employing PatternMaster (Courtesy of Sanders Prototype, Inc.)

4.3 Z-Corporation Z402 3D printer (Three-Dimensional Printing)

The Three-Dimensional Printing (3DP) process was invented and patented by the Massachusetts Institute of Technology, and licensed to Z Corporation in 1994. The first system, the Z402 (Figure 4.8), was commercialised in 1996. The Z402 system is the fastest RP machine on the market with a building speed of 25 to 50 mm per hour. The system specifications are given in Figure 4.9. The price of the Z402 system is around $50,000.

Figure 4.8 Z402 system (Courtesy of Z Corp.)

Build envelope	203 x 254 x 203 mm
Layer thickness	0.076 to 0.256 mm
Build speed	25 – 51 mm per hour
Dimensions	740 x 910 x 1070 mm
Weight	136 kg

Figure 4.9 Specifications of the Z402 system [Z Corporation, 2000].

Z Corporation offers two powder-binder systems, ZP11 and ZP100, for use with the Z402 3D Printer [Z Corporation, 2000]. ZP11 powder, the original material for the system, is composed of starch and cellulose. For additional strength, parts built from ZP11 are infiltrated with wax or a range of resins to enhance their physical properties. ZP100 is a powder material released in September 1999, which is composed primarily of plaster and delivers higher strength and detail. ZP100 is 4-5 times stronger than ZP11 but, like its predecessor, if additional strength is required parts produced from ZP100 can be infiltrated to achieve better physical properties and make the parts suitable for sanding and painting. Examples of parts produced using the Z402 system are shown in Figure 4.10.

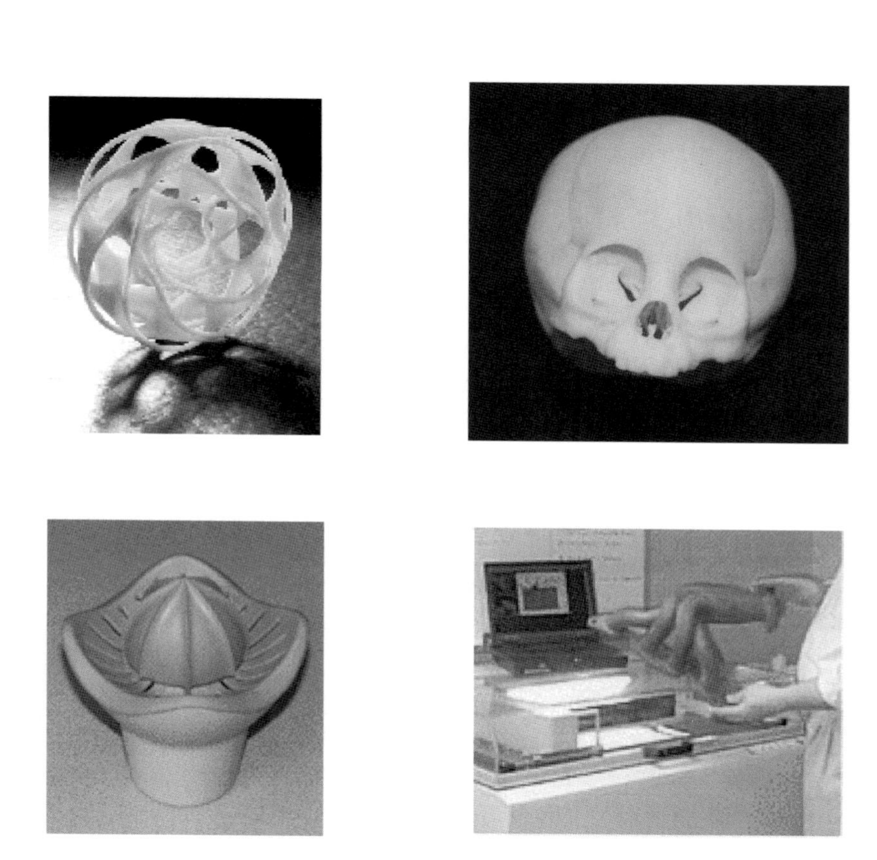

Figure 4.10* Examples of parts built using the 3DP process (Courtesy of Z Corp.)

The resolution of this process is dependent on the sizes of the binder droplets and the powder grains, the placement accuracy of the nozzle and the way that the binder diffuses through the powder due to capillary action. The layer thickness is affected by the compression of the powder due to the weight of subsequent layers. This compression is most noticeable in the centre of the part. At the base, there is no room to compact the powder. At the top of the part, there are fewer layers to cause the compaction. However, this effect is mitigated when using more densely packed powders [Soligen, 1996].

Parts do not require supports to brace overhanging features. They do however need to include a hole so that excess powder can be removed. Disadvantages of this technique are that the final parts may be fragile and porous, and it can be difficult to remove the unbound powder from any cavities. A further drawback is that the layers are raster-scanned by the printhead which leads to a stair-stepping effect in the X-Y plane as well as the build direction [Soligen, 1996].

4.4 Stratasys Genisys Xs 3D printer

The process was developed by IBM and is similar to Fused Deposition Modelling. In January 1995, Stratasys purchased this RP technology from IBM and from it developed a Concept Modeller called the Genisys 3D Printer principally for use by design offices. The latest version of the Genisys 3D printer is shown in Figure 4.11 and a summary of its specifications given in Figure 4.12.

Figure 4.11 Genisys Xs 3D printer (Courtesy of Stratasys)

Build envelope	305x203x203 mm
Speed	101 mm/second (4 in/second)
Reduction/Enlargement	Print 3D Models to meet your needs with selectable scaling
Functionality	Automatic operation; nest multiple parts in the build envelope
Compatibility	Windows NT, Sun Microsystem, Hewlett-Packard and Silicon Graphics Workstations
Material	Durable polyester
Accuracy	± 0.356mm
Size and Weight	914(w) x 816(d) x 737 (h) mm, 95kg
Power Requirements	220-240 VAC, 50/60 Hz, 6A or 110-120 VAC, 60Hz, 12A
Heat Emission	Ambient Build Temperature
Noise Emission	62 dba operating

Figure 4.12 Technical specifications of Genisys Xs [Stratasys, 2000a]

The Genisys 3D Printer builds models by extruding a bead of polyester compound through a computer-controlled pump. Material is loaded in a series of cassettes. Each cassette holds 50 rectangular wafers. A stapling mechanism feeds the wafers into a pressurised, heated channel that supplies the material to a viscosity pump [Stratasys, 2000b]. To ensure accuracy, the material is extruded through a 0.3302 mm diameter orifice at a controlled rate. Parts are built on a thermally controlled metallic substrate that rests on a table. As each layer is extruded, it bonds to the previous layer and solidifies. The pump head, table and gantry move in the X, Y and Z axes, respectively. The gantry is employed to reposition the printing nozzle in the Z direction to build subsequent layers. Supports for the models are built from the same polyester material. The system creates perforations where supports adjoin the model, making it easy to snap them off. The Genisys models often require fewer supports than those of other systems thanks to precise pump control. A technique called "bridging" allows the material to be extruded across a distance without supports [Stratasys, 2000b]. Thin perimeter walls are created, and the pump head fills in the area, creating a flat surface between the walls. This technique reduces build time and maintains a good surface finish.

4.5 JP System 5

The JP System 5 is patented by Schroff Development Corporation. It is an inexpensive RP system that requires only a personal computer and a cutting device. The system produces prototypes using paper sheets and so the material cost is very low. Two versions of the system are available: the Standard System, and the Premier System. The Standard System uses a Roland cutter with 216 x 279 mm adhesive paper. The Premier System employs a Graphtec cutter with 432 x 559 mm paper. This RP system is ideally suited for education but also can be used in other applications.

The fabrication of models employing the JP System 5 includes the following steps (Figure 4.13) [Schroff, 2000]:

1. A proprietary software package is used to slice the model into a series of cross sections.

2. The software then converts the slices to a HPGL plot file format which is sent to a plotter.

3. The sheets are layered by aligning specially cut orientation holes.

4. The sheets are positioned on a registration board.

5. An adhesive is sprayed on the first sheet.

6. The support backing is removed exposing the adhesive layer. Each subsequent sheet adheres to the exposed adhesive surfaces of the previous sheet.

7. Sections formed by the nesting operation are cut.

8. The final model is assembled.

Examples of parts built using JP System 5 are shown in Figure 4.14.

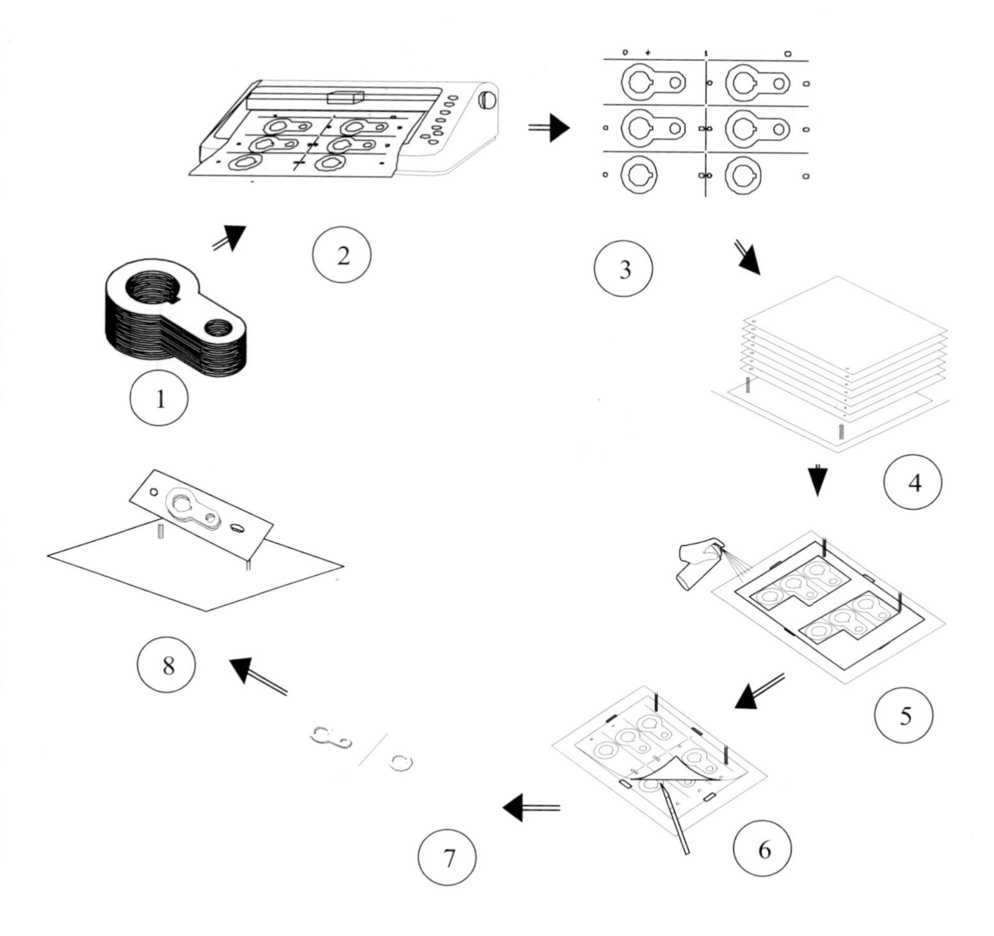

Figure 4.13 Process steps of the JP System 5 (Courtesy of Schroff Development Corp.)

Figure 4.14 Models produced using the JP System 5
(Courtesy of Schroff Development Corp.)

4.6 Objet Quadra System

Objet Geometries Ltd. was founded in 1998 to develop a RP system that employs state-of-the-art ink-jet technology and a proprietary photopolymer. The company will ship its first system called the Objet Quadra in 2001. The pre-production version of the system is shown in Figure 4.15 and its technical specifications given in Figure 4.16. The price of Objet Quadra is around $39,000 [Objet, 2000].

The Objet Quadra process employs 1536 nozzles to build parts by depositing layers of photo sensitive resin that are then fully cured, layer-by-layer, using two UV lights. As previously mentioned, models produced by the system do not require post-curing. The Objet system prints with a resolution of 600 dpi and a layer thickness of 20 μm. Only one photopolymer is currently available for building models but other materials are under development. To support overhanging areas and undercuts Objet deposits a second material which can be separated easily from the model without leaving any blemishes [Objet, 2000].

Figure 4.15 The Objet Quadra system (Courtesy of Objet Geometries)

Build envelope	270 x 320 x 200 mm
Dimensions	1290 x 890 x 870 mm
Input Format	STL file
Print Resolution	X - 600 dpi Y - 600 dpi Z - 1200 and 900(h) dpi
Building Material	Photopolymer
Communication	LAN - TCP/IP
Power Requirements	220 VAC 60 Hz/115 VAC 50 Hz. 3 kW
Operational Environment	Temperature 18° to 28° C Humidity 30% to 80%
Machine Weight	Approx. 220 kg
Computer Requirements	PC Pentium III

Figure 4.16 Technical specifications of the Objet Quadra [Objet, 2000]

The material delivery sub-system is cartridge-based for easy installation and replacement. The UV lamps are standard off-the-shelf items, costing less than $75 and lasting up to 1000 hours.

Examples of models produced using the Objet Quadra are shown in Figure 4.17.

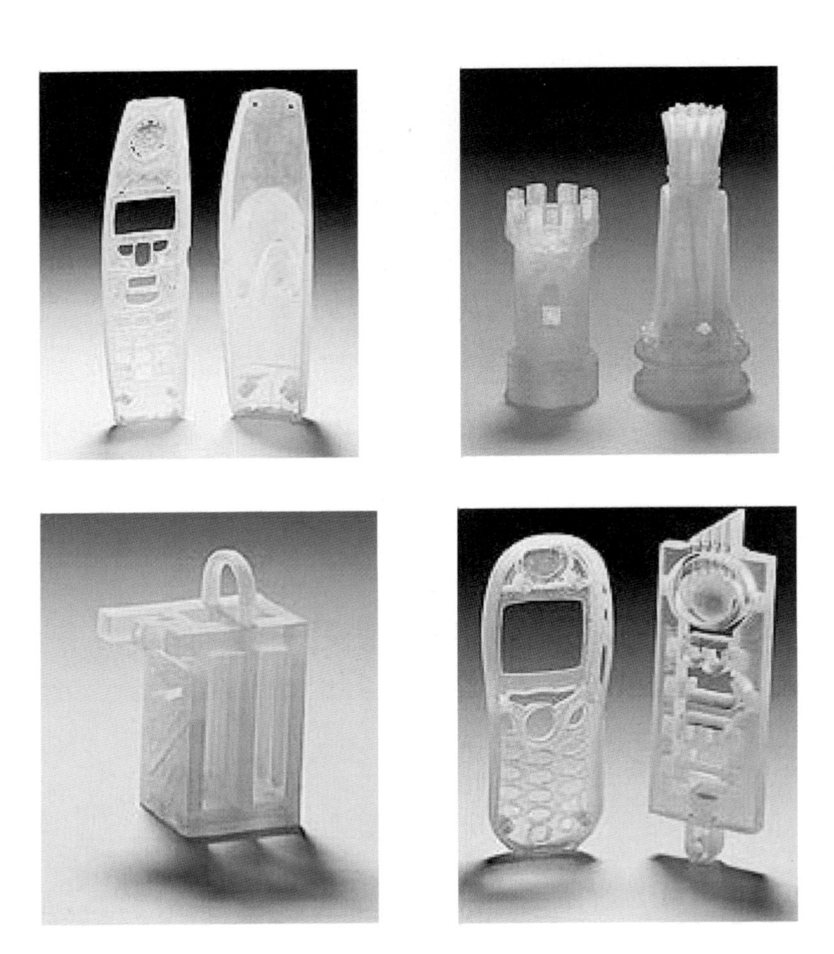

Figure 4.17* Example of models produced using the Objet Quadra (Courtesy of Objet Geometry)

4.7 Summary

Concept Modellers are relatively inexpensive RP systems that offer cost-effective solutions for fabricating physical models at any design stage. A feature of these systems is that they can be installed easily in a design office without the need for special facilities. It is expected that Concept Modellers will become a common tool in many design offices.

References

Objet Web page (2000) **Objet Geometries Ltd.**, Rehovot, Israel, http://clients.tia.co.il/objet/inner/products.html.

Sanders Web page (2000) **Sanders Prototype, Inc.**, 316 Daniel Webster Highway Merrimack, New Hampshire 03054-4115, USA, http://www.sanders-prototype.com.

Solingen Web page (2000) **Soligen Technologies**, 19408 Londelius Street, Northridge CA 91324, http://www.partsnow.com/about/06.shtml.

Stratasys Web page (2000a) **Stratasys, Inc.**, 14950 Martin Drive, Eden Prairie, MN 55344-2020 USA, http://www.stratasys.com/.

Stratasys Genisys (2000b) The 3D Printer User Manual**, Stratasys, Inc.**, 14950 Martin Drive, Eden Prairie, MN 55344-2020, USA.

Schroff Development Corporation (2000) JP System 5 Web page, **Schroff Development Corporation**, Mission KS 66222, USA, http://www.JPSYSTEM5.com/jpsystem5/.

3D Systems Press Release (1998) ThermoJet, **3D Systems**, Worldwide Corporation HQ, 26081 Avenue Hall, Valencia, California, USA.

3D Systems Web page (2000) **3D Systems**, Worldwide Corporation HQ, 26081 Avenue Hall, Valencia, California, USA, http://www.3dsystems.com/.

Z Corporation Web page (2000) **Z Corporation**, 20 North Avenue, Burlington, MA 01803, USA, http://www.zcorp.com/.

Chapter 5 Applications of Rapid Prototyping Technology

Rapid Prototyping (RP) is a continuously evolving technology. RP models are becoming widely used in many industrial sectors. Initially conceived for design approval and part verification, RP now meets the need for a wide range of applications from building test prototypes with material properties close to those of production parts to fabricating models for art and medical applications. In order to satisfy the specific requirements of a growing number of new applications, special software tools, build techniques and materials have been developed. This chapter discusses the use of RP in five different application areas: building functional prototypes, patterns for castings, medical models, artworks and models for engineering analysis. In addition, the chapter outlines the technological capabilities of RP processes in the context of each particular application and discusses specific issues relating to the efficient integration of these techniques into existing manufacturing routes.

5.1 Functional Models

There are a number of RP technologies that now meet the need for building functional prototypes with material properties close to those of production parts. One of the RP processes that is widely used for producing models for functional tests is SLS. Initially, four Nylon-based materials (Standard Nylon, Fine Nylon, Fine Nylon Medical Grade, Nylon Composite) were available commercially for this process. In 1999, these four materials were replaced by two new materials, DuraForm PA and glass reinforced DuraForm GF. DuraForm prototypes can be relatively easily finished to a smooth appearance [DTM, 2000]. The production of Nylon parts is generally cost effective when a small number (1-5) of parts is required. Before the introduction of Duraform PA, a Nylon Composite known as the ProtoForm composite was used widely for producing functional parts. A case study discussing some accuracy aspects of building functional models in ProtoForm is presented below. Also, this case study addresses some general technological issues regarding the fabrication of functional models in Nylon-based materials.

ProtoForm is a blend of 50% by weight Nylon powder with a mean particle size of 50μm and 50% by weight spherical glass beads with an average diameter of 35μm. This SLS glass-filled Nylon can be processed to near full density (Figure 5.1) and has a high modulus and good heat and chemical resistance (Figure 5.2)

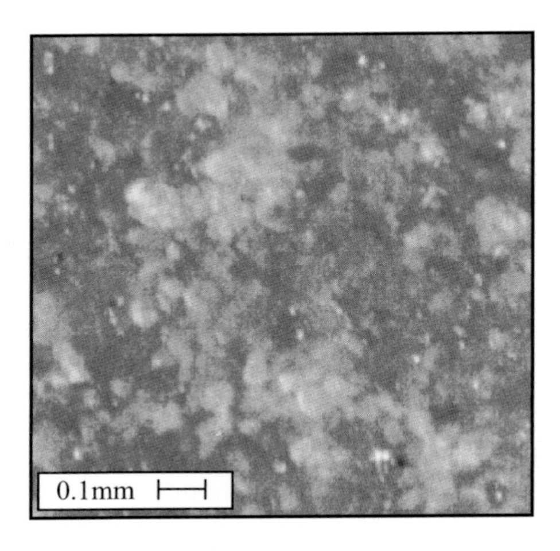

Figure 5.1 Optical micrograph of a ProtoForm Composite part surface

The housing in Figure 5.3 is a test part and is built in ProtoForm Composite because it is required to withstand harsh testing conditions including temperatures of about 100°C. As a base part for mounting precision components, it has to keep its dimensions within close limits.

Due to its overall dimensions (190x50x250 mm), the part was constructed vertically to fit within the build area (∅305x410 mm) of the DTM Sinterstation 2000. To speed up cooling, the downdraft (downward forcing of gas through the powder) capability of the machine was utilised. However, the geometry of the housing prevented the downdraft, leaving a hot area inside the part and causing post-build warping of the walls.

Tensile Strength	MPa	49
Tensile Modulus	MPa	2800
Tensile Failure Strain	%	6
Flexural Modulus	MPa	4300
Unnotched Izod	J/m	440
Notched Izod	J/m	68
DTUL (0.45Mpa)	°C	188
DTUL (1.8Mpa)	°C	134
Surface Roughness, Ra (As processed)	μm	15
Surface Roughness, Ra (Finished)	μm	4

Figure 5.2 ProtoForm Composite properties [Beaman et al., 1997]

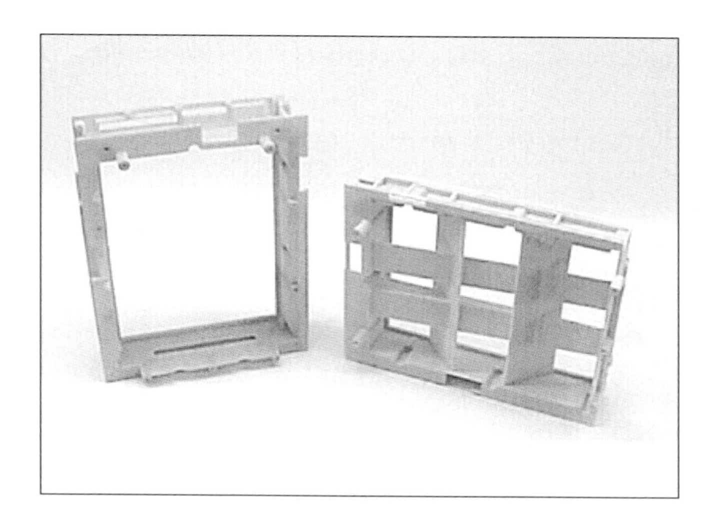

Figure 5.3 Composite Nylon housing: Without ribs (left) – With ribs (right)

The first part manufactured suffered from much distortion: there was vertical growth and "wash out" (loss of definition and rounding of edges) on the downward facing surfaces and the external dimensions of the sidewalls varied by more than 1mm. This problem was solved by making the wall thickness uniform and reducing it to 2mm. Furthermore, 2mm non-functional ribs were added across the housing to stiffen it. Two ribs were positioned vertically and two others horizontally as shown in Figure 5.3. The number and size of the ribs were determined from experience to constrain post process distortion in the X and Y directions without adding too much build time. The ribs were also located so that they could easily be removed by machining after completing the build. Subsequently manufactured parts had much better dimensional accuracy. The main functional dimensions were measured but no form or geometrical accuracy measurements were taken. Figure 5.4 shows the distribution of errors in the dimensions of the housings with and without ribs. The error in 90% of all dimensions for the modified part was between +0.35 and –0.31mm.

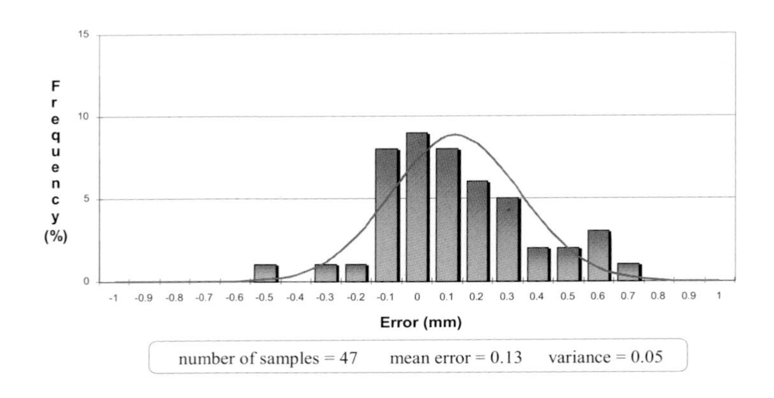

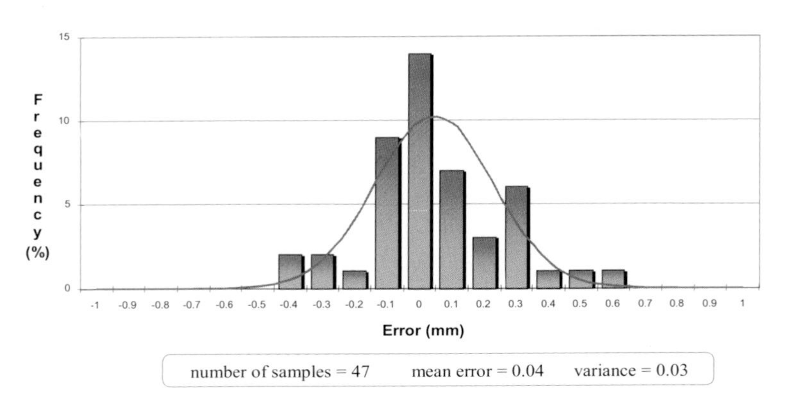

Figure 5.4 Composite Nylon housing (without and with ribs): Error distribution

To evaluate the influence of the ribs, the accuracy in the build direction (Z direction) and X-Y plane was studied. The results presented in Figure 5.5 show that the ribs improved the accuracy consistently but had different effects in each direction. The largest changes were along the X axis where both mean error and variance were considerably reduced, equalling the results recorded for the Y axis without ribs. This was expected because the X direction was the most affected by distortion. As the majority of the dimensions in the Y direction were smaller than 50mm, the addition of ribs gave mixed results, decreasing the variance but also raising the mean error. Improvements were also achieved in the Z direction where the mean error was significantly reduced. This can be attributed to the reduced post-process distortion due to the added ribs. Another indicator of the quality of a part produced by RP is the international tolerance (IT) grade established by the ISO-ANSI standards [Ippolito et al., 1995]. The cumulative IT distribution for the modified Nylon housing is given in Figure 5.6 and shows very good results with 90% of the measurements at IT 13 or better.

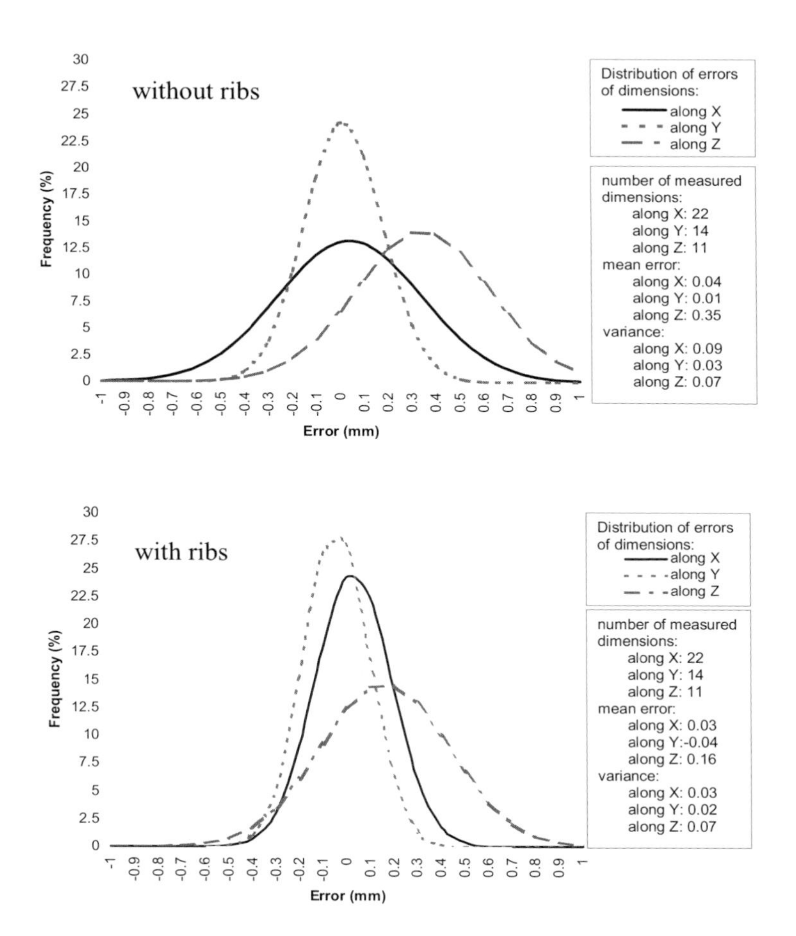

Figure 5.5 Composite Nylon housing (without and with ribs): Dimensional error distribution along X,Y and Z axes

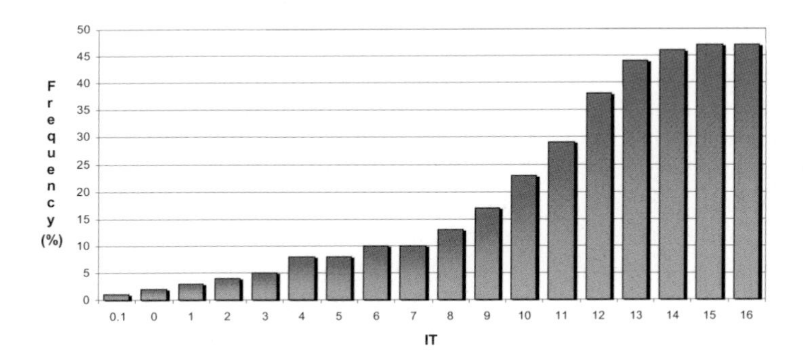

Figure 5.6 Composite Nylon housing (with ribs): Cumulative IT distribution

5.2 Patterns for Investment and Vacuum Casting

RP technologies are widely used for building patterns for investment and vacuum casting. For example, models built employing SLA, SLS and FDM can be used as patterns for both casting processes. A case study is presented below that discusses some accuracy aspects of producing SLS patterns and also addresses general issues regarding the technological capabilities of the process.

Two SLS materials are currently available for producing casting patterns, CastForm and TrueForm. In this case study, TrueForm, which is an acrylic-based powder of spherical particles with a mean diameter of 30μm, is used to build casting patterns. It is processed at relatively low temperatures compared with nylon-based materials which limits shrinkage to 0.6% [Van de Crommert et al., 1997] and is the preferred material for making parts with good accuracy but moderate strength. The density of TrueForm parts can vary from 70 to 90% depending on build parameters and they can be polished to a mirror-like finish. Dense parts are used as patterns for vacuum casting while rather porous parts are better suited for investment casting; unlike dense models they do not expand to cause shell cracking during the burning out of the patterns [McAlea et al., 1996] (Figure 5.7).

Figure 5.8 shows a TO BS L99 aluminium housing (195x145x250mm) made by investment casting from a TrueForm pattern.

The distribution of actual dimensional errors for all axes is plotted in Figure 5.9. It can be seen that 90% of the errors are between +0.7 and -0.6mm. Figure 5.10 shows the distributions for individual axes to reveal directional influence. The accuracy of the part is highly dependent on its size, the largest errors being found on the largest dimensions. To remove the effect of size, Figure 5.11 depicts the spread of errors for individual axes as percentages of the dimensions measured. Figures 5.12 shows that 90% of the errors are at IT 13 or better.

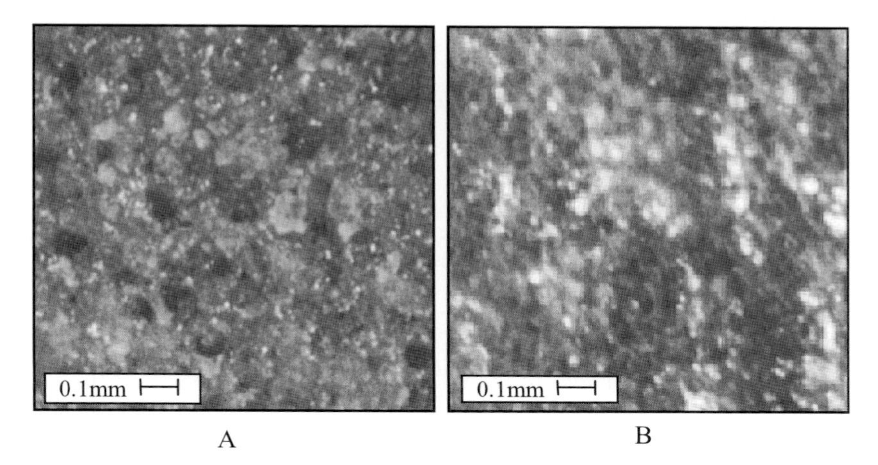

Figure 5.7 Optical micrograph of a TrueForm part surface (A: Porous, B: Dense)

Figure 5.8 TrueForm housing pattern (right) and aluminum investment casting (left)

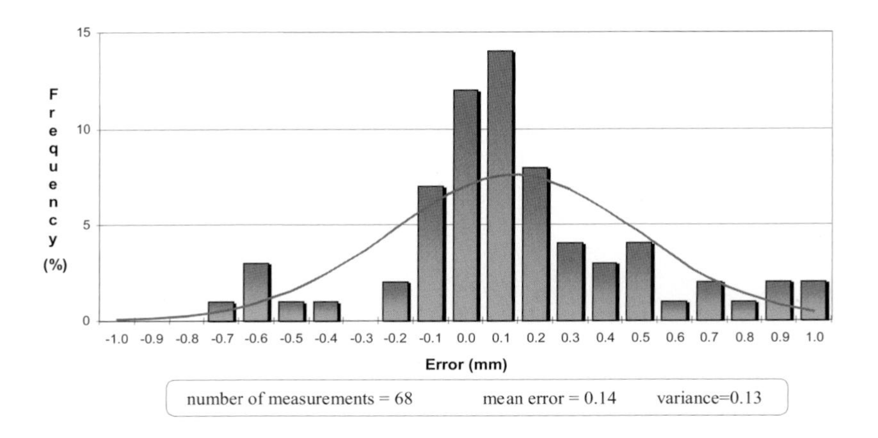

Figure 5.9 TrueForm housing: Distribution of all errors

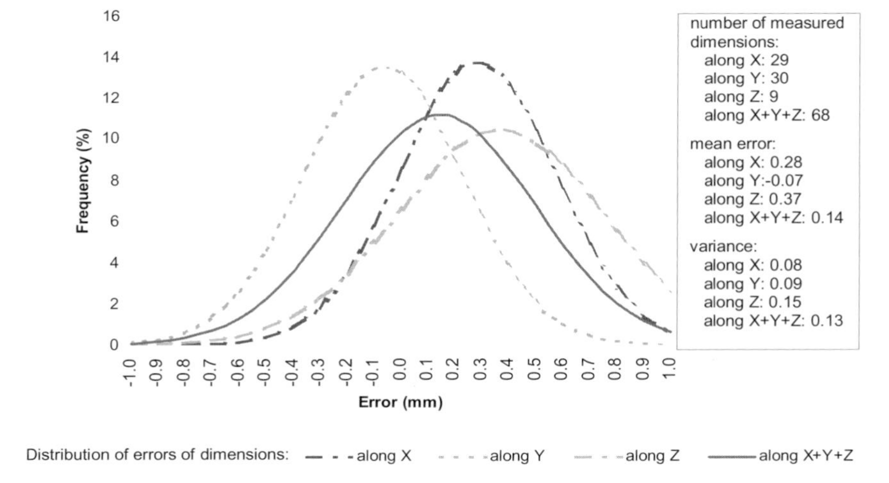

Figure 5.10 TrueForm housing: Distribution of errors along different directions

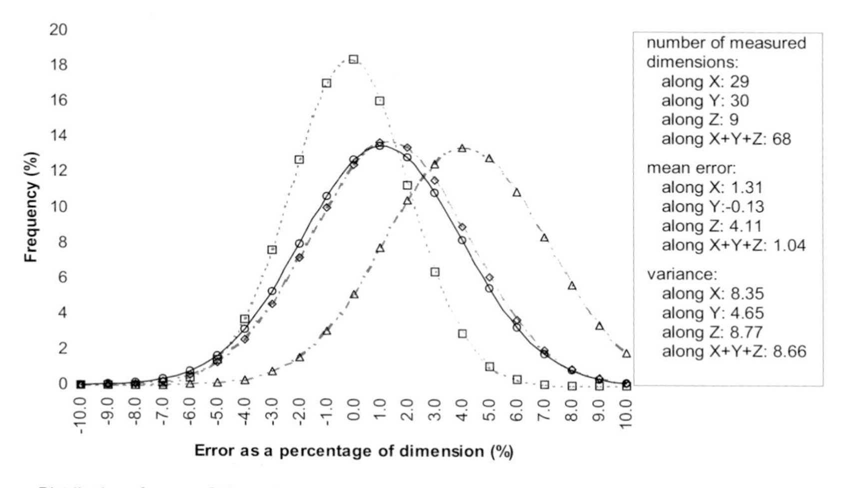

Figure 5.11 TrueForm housing: Distribution of percentage errors along different directions

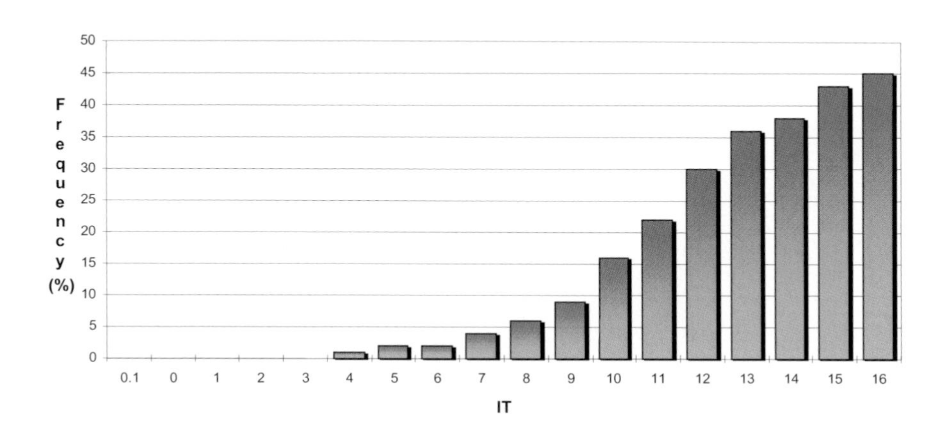

Figure 5.12 TrueForm housing: Cumulative IT distribution

TrueForm behaves rather like an injection moulded plastic and thick sections may be subjected to sinking or sagging. Part orientation must, where possible, be selected to prevent sagging. This may not always be practicable and in such cases shelling of the model (i.e. converting a solid model into a hollow part) can significantly reduce part distortion (Figures 5.13, 5.14 and 5.15). TrueForm normally gives good accuracy but, in this case, the part was oriented such that most of the dimensions measured were in the Z direction which, as mentioned previously, is the direction most affected by inherent error sources. This explains why the achieved accuracy was not better.

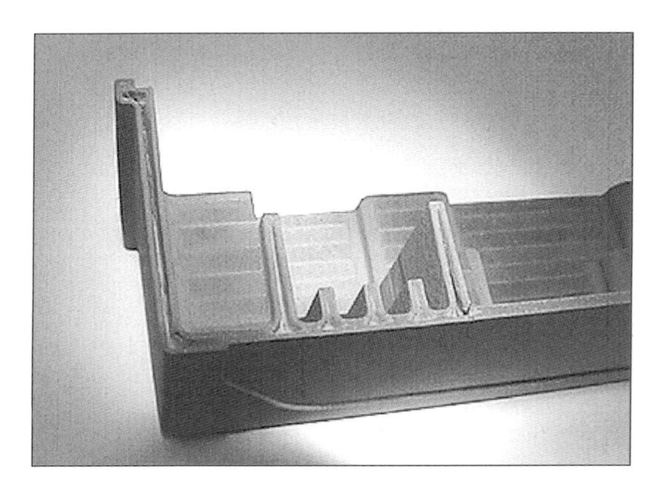

Figure 5.13 Cross section of a hollow TrueForm part

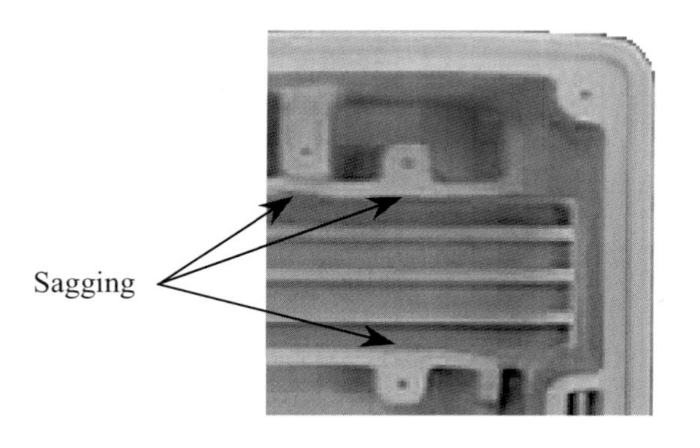

Figure 5.14 Sagging due to non-constant wall thickness

Figure 5.15 Improvement of the accuracy on a hollow part

If a larger SLS machine such as the Sinterstation 2500 had been used this would have allowed the part to be oriented horizontally in the build area. In this way, the accuracy would certainly have been much closer to ±0.125mm which is the accuracy quoted for the TrueForm material by the machine manufacturer. However, even if some dimensions were out of the required general tolerance (±0.125mm) the aluminium castings were fully satisfactory as any deviations were able to be corrected when some of the features were machine-finished afterwards. TrueForm patterns become cost effective when a small number of parts, say up to 50, of complex design are required and the cost of a mould for wax patterns is prohibitive.

5.3 Medical Models

RP technologies are applied in the medical domain for building models that provide visual and tactile information. In particular, RP models can be employed in the following medical applications [Anatomics, 2000; D'Urso, 1998; D'Urso, 1999a; D'Urso, 1999b; Materialise, 2000].

- *Operation planning.* Using real size RP models of patients' pathologic regions, surgeons can much more easily understand physical problems and gain a better insight into the operations to be performed. RP models can also assist surgeons in communicating the proposed surgical procedures to the patients.

- *Surgery rehearsal.* RP models offer unique opportunities for surgeons and surgical teams to rehearse complex operations using the same techniques and tools as during actual surgery. Potentially, such rehearsals can lead to changes in surgical procedures and significantly reduce risk.

- *Training.* RP models of specimens of unusual medical deformities can be built to facilitate the training of student surgeons and radiologists. Such models can also be employed for student examinations.

- *Prosthesis design.* RP models can be used to fabricate master patterns which are then replicated using a bio-compatible plastic material. Implants produced in this way are much more accurate and cost effective than those produced employing conventional techniques.

The building of RP models of anatomical structures involves the following steps:

1. *Data acquisition with medical equipment.* Conventional 3D medical scanners (CT Scans, MRI Scans, 3D Ultrasound) are employed to capture a sequence of images of a particular anatomic structure. In Figure 5.16, a CT scanner is shown together with a captured image.

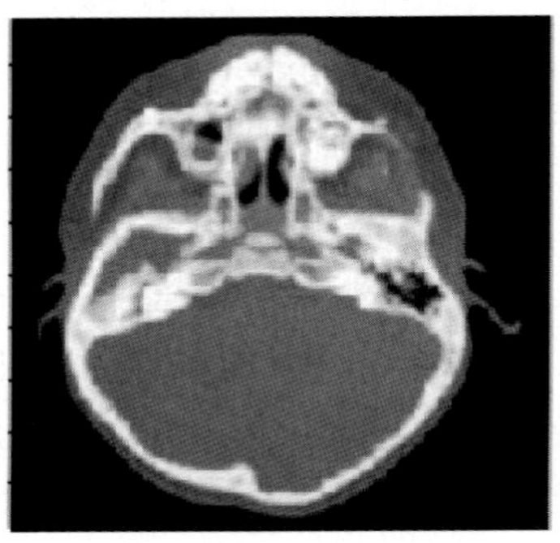

Figure 5.16 A CT scanner and a captured image (Courtesy of Materialise)

2. *Generation of STL files from the scan data.* Interactive software tools exist for segmentation of scanned images and generation of STL files. For example, the MIMICS software (Figure 5.17) developed by Materialise enables users to control and correct the segmentation of scanned images [Materialise, 2000]. The segmentation volume is defined by all pixels with a grey value higher than a predefined threshold. Also, the software allows a definition of the segmentation volume with a grey value in-between two threshold values. This technique can be employed for segmentation of soft tissue in CT images or for segmentation of several structures in MR images.

3. *Building RP models from the generated STL files.* Any RP technology can be employed for building medical models. In addition to the general-purpose materials for each RP technology, special materials have been developed for medical applications. For example, a Fine Nylon Medical Grade for the SLS process and a resin, Stereocol, for the SLA process are dedicated to medical applications. Fine Nylon MG is a modification of the fine nylon that can be sterilised in an autoclave. The Stereocol resin is not as accurate as an epoxy resin but offers several advantages for medical applications. For instance, this resin can be coloured by hatching up certain areas of the models a second time with an UV laser. Because SLA models are translucent, the coloured regions

can be viewed within the parts. Figure 5.18 shows a medical model built using Stereocol.

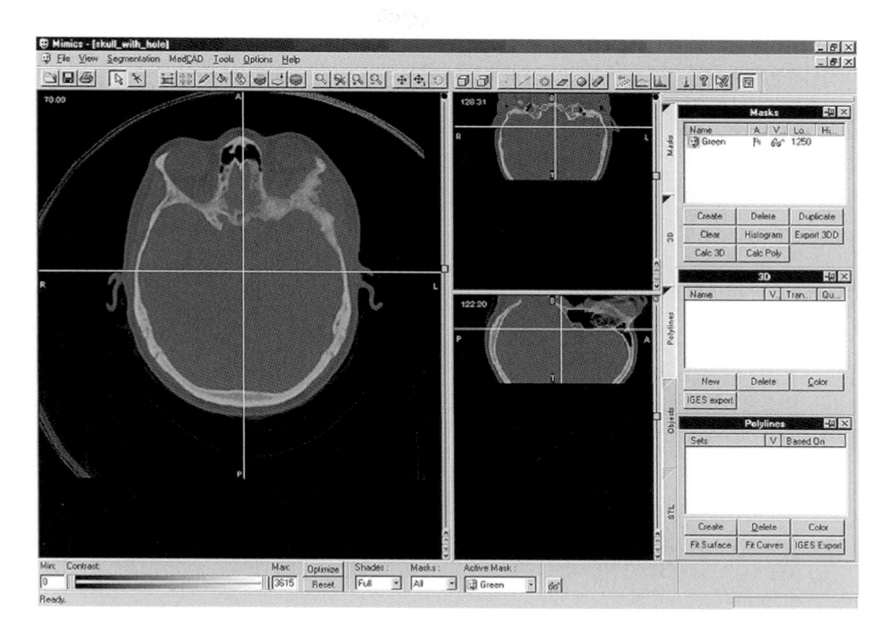

Figure 5.17* User interface in the MIMICS software (Courtesy of Materialise)

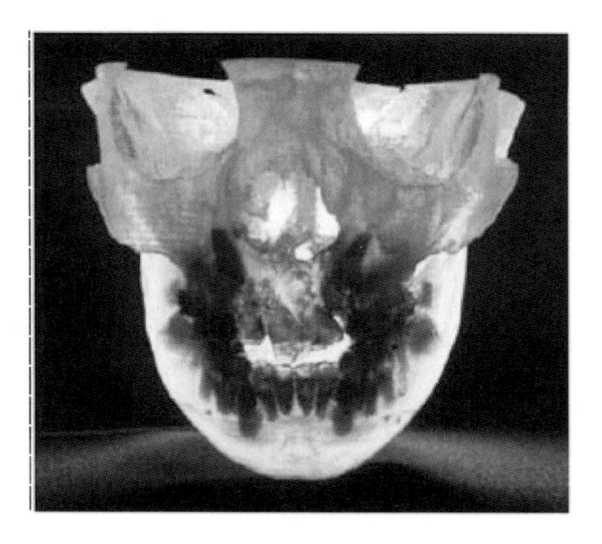

Figure 5.18* A medical model built using the Stereocol resin (Courtesy of Materialise)

Two case studies are described below that demonstrate the use of RP models in the medical domain. In the first case study reported by Anatomics™, Queensland, Australia [Anatomics, 2000], two SLA medical models were built for a patient suffering from a secondary carcinoma of the right superior orbital margin and the adjacent frontal bone. The first model was used to plan the resection of the cancerous bone and also as an operation reference and patient consent tool. The SLA model can be cut employing the same surgical tools as those used for bone resectioning. A resectioned template was created in plastic following the surgeon's desired resection line. The fabricated plastic template was placed over the model to check the match with the surgeon's resection line (Figure 5.19).

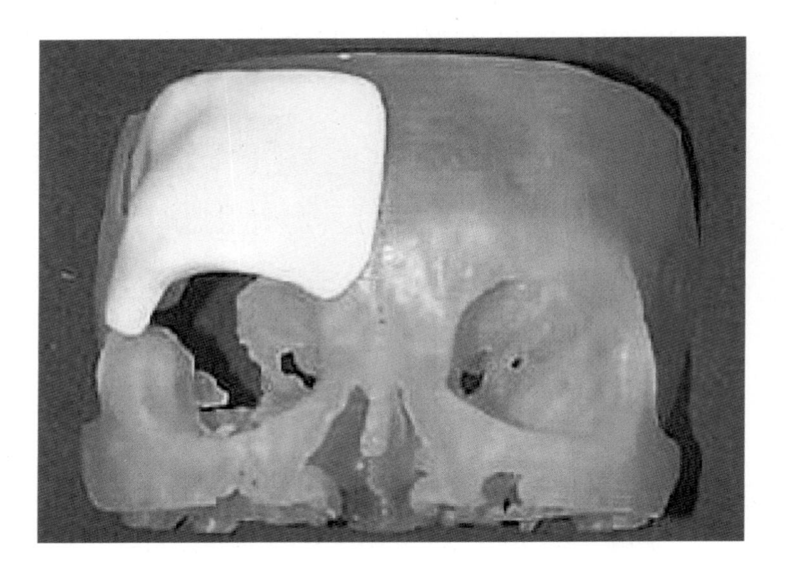

Figure 5.19* The SLA model with the resection template (Courtesy of the British Journal of Neurosurgery)

The second model was then employed to construct a custom acrylic implant (Figure 5.20). To achieve this, the unaffected left superior orbital margin was mirrored across to assist the design of the implant. The resection template and the custom implant were prepared for the operation by gas sterilisation. The template was then placed onto the lesion and the resection line traced out and the bone cut away. Finally, the implant was inserted into the deficit.

The operation was a complete success. The surgeon was fully satisfied with the quality and the cost of utilising RP models.

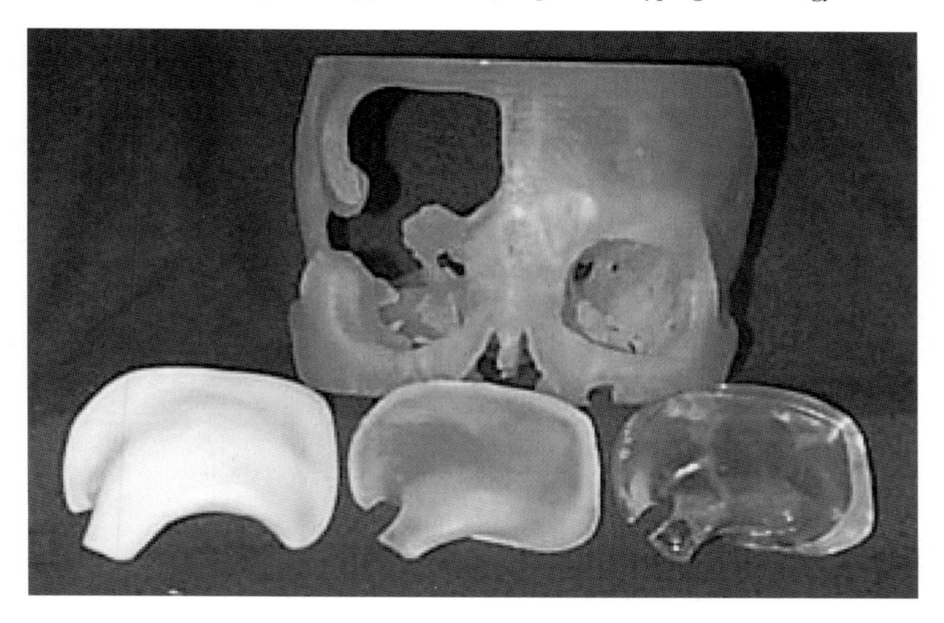

Figure 5.20* The SLA model together with the template and the implant (Courtesy of the British Journal of Neurosurgery)

The second case study reported by Materialise [Materialise, 2000] demonstrates the use of RP technology for the fabrication of an obturator prosthesis for an oncologic patient. CT scanned data was employed to build an RP tool for direct moulding of an implant in bio-compatible silicone. Figure 5.21 shows the different steps involved to complete this medical intervention. The built prosthesis fitted much better than those produced by traditional techniques. The RP prosthesis is much more accurate and considerably reduces the damage caused to sensitive and vulnerable surrounding tissues during surgery.

Another growing application area for RP technologies is in art and design. Through building RP models, artists can experiment with complex artworks which support and enhance their creativity. Initially, the high cost of RP models meant strict limits on the size of the models. However, recently, with the introduction of Concept Modellers, it has become cost effective to employ RP techniques in many artistic applications. Taking into account the accuracy of art models and the RP materials available, the technological capabilities of Concept Modellers are more than adequate for the majority of art applications.

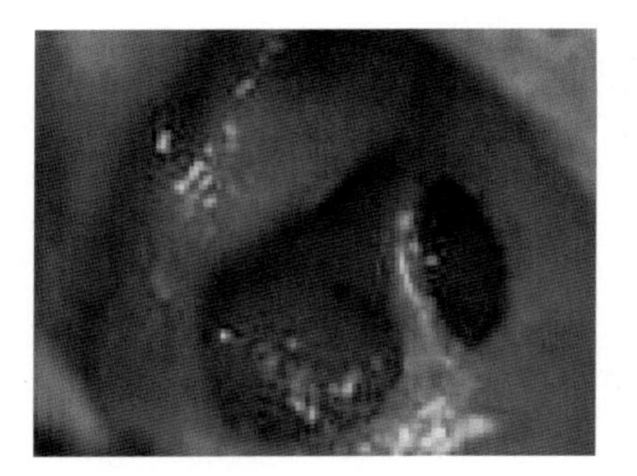

A: A hole resulted from the irradiation of a tumor in the mouth cavity. This hole has to be filled by an implant to allow the patient to breathe and eat normally.

B: The soft tissue surrounding the cavity was modelled by CT scanning. This model was used to design a tool for direct moulding of the implant.

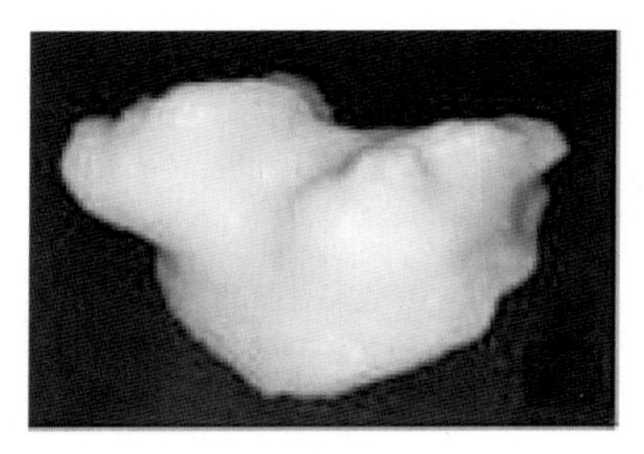

C: A silicon implant was moulded from the tool.

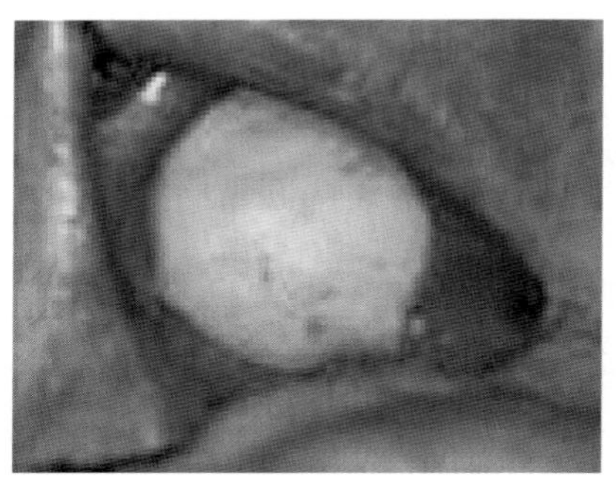

D: Without surgery, the deformable silicon prosthesis was implanted. Magnets were used to fix the prosthesis to a hard dental implant.

Figure 5.21* Fabrication of obturator prosthesis using RP techniques. Case study presented by Dr L.L. Visch from Daniel den Hoed Kliniek Rotterdam (Courtesy of Materialise)

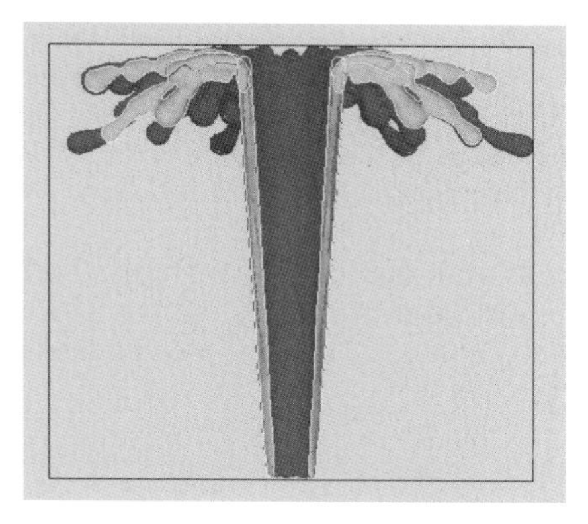

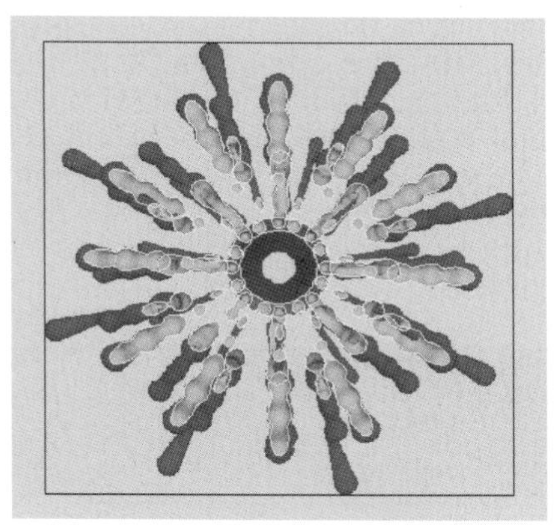

Figure 5.22* Cross-sections of the 3D model of a water splash (Courtesy of the CALM project)

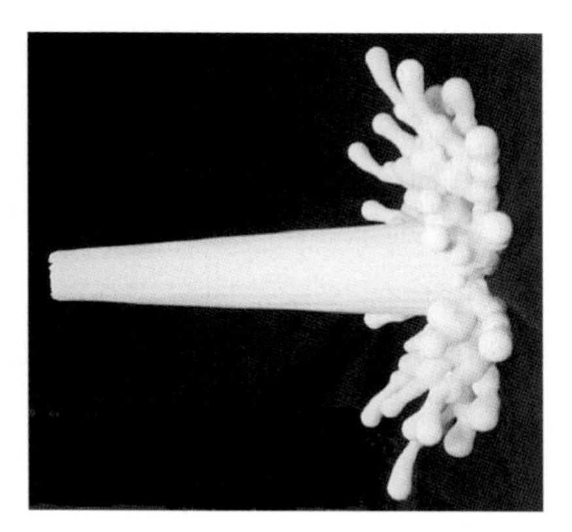

Figure 5.23 SLS model representing a water splash (Courtesy of the CALM project)

Three examples are described below that demonstrate the use of RP techniques in art. These were parts of work conducted within the CALM (Creating Art with Layer Manufacture) project [CALM, 1998] that was supported by the Higher Education Funding Council for England as part of an initiative to promote the use of IT within the Art and Design community in UK higher education.

The first example is an artwork created by M. Harris. The art model represents a splash spanning the inside of a plexiglass vitrine (Figure 5.22). In its final installation, the RP model (Figure 5.23) will be incorporated into a plexi-box exactly the width of the splash itself.

The second example is a cybersculpture created by K. Brown. The artwork represents an artefact (Figure 5.24) which cannot be built using any conventional methods. The initial intentions of the author were to produce a RP pattern and then cast it in bronze. However, after seeing the SLS model (Figure 5.25), he immediately recognised that "the material (Duraform) in conjunction with the lace-like Moiré surface patterns makes for an immaculate object in itself. The combined integrity of the concept, process, material, colour and form (including the strata of layers that run through it) add up to something the like of which I have not experienced before" [CALM, 1998].

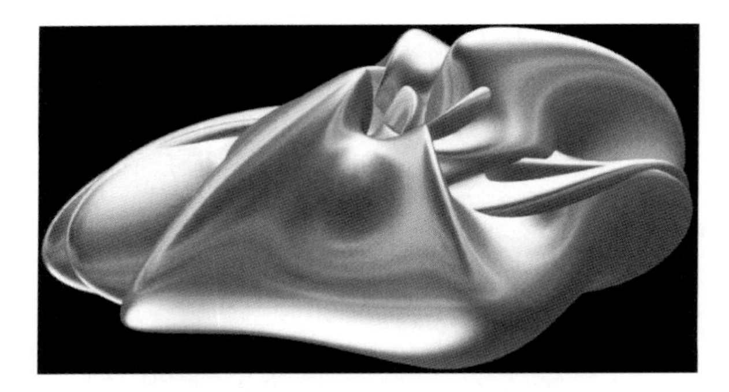

Figure 5.24* 3D shaded image and a cross-section of the cybersculpture (Courtesy of the CALM project)

Figure 5.25 SLS model of a cybersculpture
(Courtesy of the CALM project)

The third example of a work of art produced employing RP techniques is a bracelet designed by M. Woolner and T. Cook [CALM, 1998]. The bracelet is designed in such a way that it can be produced only by using RP technologies. The artefact is a complex trajectory extrusion starting at one end as an ellipse with a star-shaped internal form and finishing at the other end with a star-shaped section having an ellipse as an internal form. The RP model was built using a SLA system. Both models are shown in Figure 5.26 The authors' intention was to cast the bracelet in silver after making minor modifications to the design.

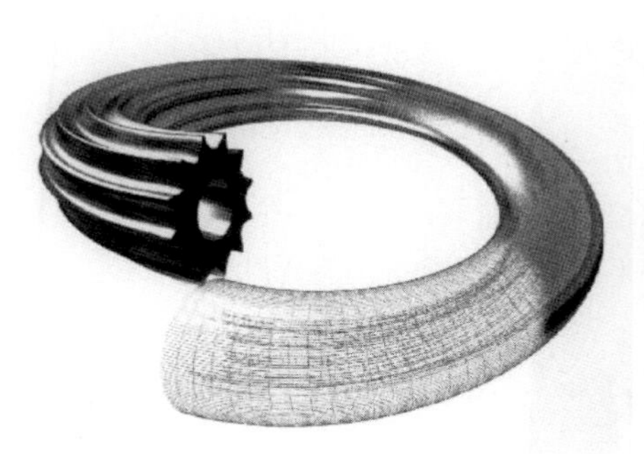

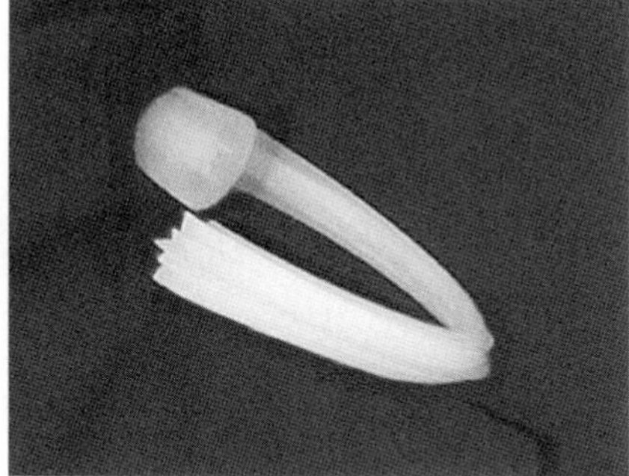

Figure 5.26 3D CAD model and the SLA art prototype
(Courtesy of the CALM project)

5.5 Engineering Analysis Models

Computer Aided Engineering (CAE) analysis is an integral part of Time-Compression technologies. Various software tools exist, mainly based on Finite Element Analysis (FEA), to speed up the development of new products by initiating design optimization before physical prototypes are available. However, the creation of accurate FEA models for complex engineering objects sometimes requires significant amounts of time and effort [Gartzen et al., 1998; Jacobs, 1996; Raymond and Thomas, 1998]. By employing RP techniques it is possible to begin test programmes on physical models much earlier and complement the CAE data. Four applications of RP models for engineering analysis are described below.

1. *Visualisation of Flow Patterns.* For example, SLA models were used to optimise the cross-flow jacket of a V6 high-performance racing engine (Figure 5.27) [Jacobs, 1996]. 60 sensors were installed in the model to monitor local flow temperature and pressure conditions. The coolant flow patterns were visualised by accurately injecting very small air bubbles. The flow patterns were recorded by high-speed video. The analysis conducted provided valuable data about stagnation zones and insufficiently cooled sections. This data allowed the critical sections to be redesigned and SLA models of the modified components were produced. Each design iteration took one week. This enabled the testing of flow channel variations in a very short time. The accuracy and surface quality of the SLA models are more than adequate to reproduce complex flow behaviours.

Figure 5.27 Assembly of the cross-flow water jacket of a V6 high-performance racing engine [Jacobs, 1996]

2. *Thermo Elastic Tension Analysis (THESA).* By employing the THESA method [Gartzen et al., 1998], RP models of real parts can be used on test rigs for structural analysis. This method allows temperature changes in the test parts to be directly correlated to the load. The effect of a particular load on the temperature patterns is analysed using thermal imaging. For example, SLS models built in glass-filled nylon were utilised to optimise the design of a highly loaded wheel rim of a sports car [Gartzen et al., 1998]. Initially, a rim segment was produced in a number of RP materials to find the optimal material for the THESA investigation. The results showed that the temperature patterns of the glass-filled nylon model were very similar to those obtained from cast metal parts. Thermographic plots of an aluminum rim and a SLS rim are shown in Figure 5.28. This case study demonstrated that the test results obtained on SLS models using the THESA method could be correlated to the behaviour of series cast metal parts.

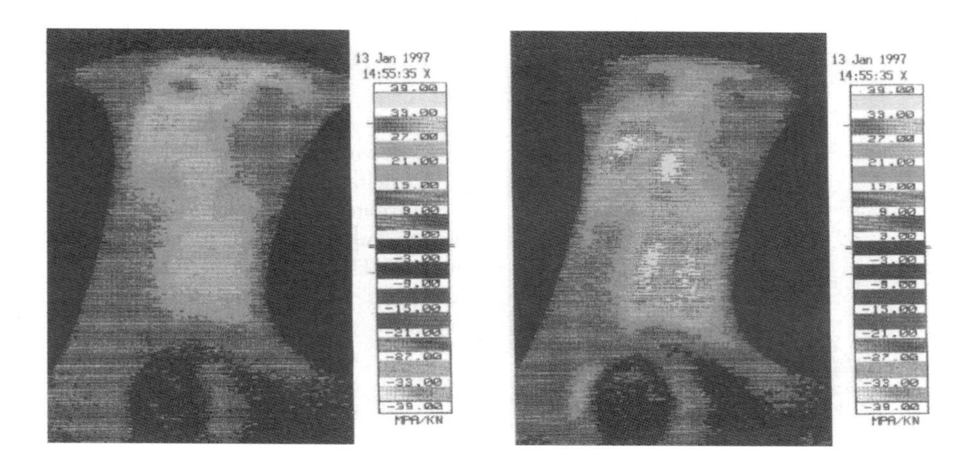

Figure 5.28* Thermographic plots of aluminium rim (left) and SLS glass-filled nylon rim (right) [Gartzen et al., 1998]

3. *Photoelastic Stress Analysis.* Photoelastic testing can be used to determine the stresses and strains within physical parts under specific conditions. This analysis is based on the temporary birefringence of a transparent material subjected to a specific load [Jacobs, 1996]. SLA models fabricated using the ACES build style (see Chapter 3) exhibit the required birefringence that can be illustrated by irradiating the test samples with polarised white and monochromatic light. Results from photoelastic analysis of SLA models can be transferred to functional metal parts by employing fundamental similarity laws. This allows predictions to be made rapidly at low cost regarding the actual stress distributions anticipated in functional parts. By using SLA models for

photoelastic stress analysis, it is also possible to "freeze" the stresses and strains by warming the loaded model to a level above the resin glass transition temperature and then gradually cooling the model back to room temperature (Figure 5.29) [Jacobs, 1996].

Figure 5.29* The frozen stress distribution for a model of an aeroengine turbine rotor [3D Systems, 1994]

4. *Fabrication of Models for Wind Tunnel Tests*. Despite recent advances in CAE tools, the aerospace and automotive industries still rely on experimental wind tunnel test data to verify the performance of new designs. RP techniques can be used to produce wind tunnel models which are not subjected to significant loads [Raymond and Thomas, 1998]. For example, the strength, accuracy and surface finish of models produced using SLA, SLS, FDM and SGC technologies are sufficient for tests of non-structurally loaded parts. In addition, SLS models produced in RapidSteel or metal models fabricated from RP patterns are adequate for lightly loaded applications.

5.6 Summary

This chapter has provided an overview of RP model application areas and of the specialised software tools available, building styles and materials used. A number of case studies have been described to assist users in identifying possible applications for RP technologies within the context of their particular needs. As RP is a constantly evolving technology, the list of application areas provided in this chapter should not be considered exclusive. There are a number of R&D centres worldwide seeking new RP applications in parallel with work on technology and materials development.

References

3D Systems Newsletter (1994) **The Edge**, Summer, 3D Systems, 26081 Avenue Hall, Valencia, California, USA.

Anatomics Case Studies (2000), **Anatomics Rty. Ltd**, Queensland, Australia, http://glacier.qmi.asn.au:80/anatomics/.

Beaman JJ, Barlow JW, Bourell DL, Crawford RH, Marcus HL and McAlea KP (1997) Solid Freeform Fabrication: A New Direction in Manufacturing, **Kluwer Academic Publishers**, Dordrecht, The Netherlands.

CALM Project Final Report (1998) University of Central Lancashire, Preston, http://www.uclan.ac.uk/clt/calm/overview.htm.

DTM White paper (2000) Functional prototyping with DuraForm and SLS, **DTM Corporation**, Austin, Texas, USA, 2000.

D'Urso PS, Atkinson RL, Lanigan MW, Earwaker WJ, Bruce IJ, Holmes A, Barker TM, Effeney DJ and Thompson RG (1998) Stereolithographic biomodelling in craniofacial surgery, **The British Journal of Plastic Surgery**, Vol. 51, 7, pp 522-530.

D'Urso PS, Atkinson RL, Weidmann MJ, Redmond MJ, Hall BI, Earwaker WJ, Thompson RG and Effeney DJ (1999a) Biomodelling of skull base tumours. **The Journal of Clinical Neuroscience,** Vol. 6, 1, pp 31-35.

D'Urso PS, Barker TM, Earwaker WJ, Bruce IJ, Atkinson RL, Lanigan MW, Arvier JF and Effeney DJ (1999b) Stereolithographic biomodelling in cranio-maxillofacial surgery: a prospective trial. **The Journal of Cranio-maxillofacial Surgery**, Vol. 27, pp 30-37.

D'Urso PS and Redmond MJ (2000), Method for the Resection of Cranal Tumours and Skill Reconstruction. **British Journal of Neurosurgery**. Vol. 4, 6, pp 555-559, http://www.tandf.co.uk, see also [Anatomics, 2000].

Gatrzen J, Lingens H, Gebhardt A and Schwarz C (1998) Optimisation using THESA, **Prototyping Technology International '98**, UK & International Press, Surrey, UK, pp 36-38.

Ippolito R, Iuliano L and Gatto A (1995) Benchmarking of Rapid Prototyping Techniques in Terms of Dimensional Accuracy and Surface Finish, **CIRP Annals**, Vol. 44, 1, pp 157-160.

Jacobs PF (1996) Stereolithography and other RP&M technologies, **Society of Manufacturing Engineers - American Society of Mechanical Engineer.**

Materialise Product Information (2000) Mimics software, **Materialise**, Leuven, Belgium, http://www.materialise.be/.

McAlea K, Lackminarayan U and Maruk P (1996) Selective Laser Sintering of Metal Molds: The Rapid Tool™ Process, **Moldin'96**, ECM Inc., Plymouth, MI .

Lightman AJ, Vanassche B, D'Urso P and Yamada S (1995) Applications of Rapid Prototyping to Surgical Planning - A Survey of Global Activities. **Proc of the 6th International Conference on Rapid Prototyping**, June 4-7, Dayton, Ohio, USA, pp 16-21.

Pham DT, Dimov SS and Lacan F (1999) Selective Laser Sintering: Applications and Technological Capabilities, **Proc. IMechE, Part B: Journal of Engineering Manufacture**, Vol. 213, pp 435-449.

Poulsen M, Lindsay C, Sullivan T and D'Urso PS (1999) Stereolithographic modelling as an aid to orbital brachytherapy. **The International Journal of Radiation Oncology, Biology and Physics**, Vol. 44, 3, pp 731-735.

Raymond NC and Thomas VJ (1998) A Comparison of Rapid Prototyping Techniques Used for Wind Tunnel Model Fabrication, **Rapid Prototyping Journal**, MCB University Press, Vol. 4, 4, pp 185-196.

Van de Crommert S, Seitz S, Esser KK and McAlea K (1997) Sand, Die and Investment Cast Parts via the SLS Selective Laser Sintering process, **DTM GmbH**, Hilden, Germany.

Chapter 6 Indirect Methods for Rapid Tool Production

As Rapid Prototyping (RP) is becoming more mature, material properties, accuracy, cost and lead-time are improving to permit it to be employed for the production of tools. Some traditional tool making methods based on the replication of models have been adapted and new techniques allowing tools to be made directly by RP have been developed. This chapter reviews indirect methods for rapid tooling (RT) that are, or shortly will be, available for production runs of up to several hundred parts in the same material as, or a material very similar to, that of the final production part. The RT methods presented below are called indirect because they use a RP pattern obtained by an appropriate RP technique as a model for mould and die making.

6.1 Role of Indirect Methods in Tool Production

In recent years, RP technologies have emerged to reduce the delays inherent in the re-iterations and fine-tuning necessary to create a high quality product. These technologies offer the capability of rapid production of three-dimensional solid objects directly from designs generated on CAD systems. Instead of several weeks, a prototype can be completed in a few days or even a few hours. Unfortunately, with RP techniques, there is only a relatively limited range of materials from which prototypes can be made. Consequently, although visualisation and dimensional/geometric verification are possible, functional testing of prototypes often is not, due to the different mechanical and thermal properties of the prototype compared to the production part [Jacobs, 1996a].

All this leads to the next step, which is for the RP industry to target tooling as a natural way to capitalise on 3D CAD modelling and RP technology. With increases in the accuracy of RP techniques, numerous processes have been developed for producing tooling from RP masters [Childs and Juster, 1994]. The most widely employed indirect RT methods are to use RP masters to make silicone room temperature-vulcanising (RTV) moulds for plastic parts and as sacrificial models for investment casting of metal parts [Dickens et al., 1995]. These processes, which are suitable for batches of 1 to 20 parts, are usually known as "soft tooling" techniques. In spite of the widening of the range of materials allowed by soft tooling, the choice is still limited and not all needs can be satisfied. Therefore, other indirect methods for tool fabrication have been developed. These new methods allow prototypes to be built using the same material and manufacturing process as the production part.

The indirect methods to be described in this chapter are a good alternative to traditional mould making techniques. These less expensive methods with shorter lead-times allow tool validation to be conducted before changes become very costly. The aim of these RT methods is to fill the gap between RP and hard tooling by enabling the production of tools capable of short prototype runs.

The broad range of indirect RT solutions makes it difficult to determine the most appropriate method for a particular project. Companies need to know all of the available processes and have a clear understanding of their strengths and weaknesses together with the comparative merits of the various materials they employ.

6.2 Metal Deposition Tools

This process involves using an RP model with a good surface finish that incorporates a draft angle and has an allowance for the shrinkage of the moulding material. The pattern is embedded along its parting line into plasticine within a chase. The sprue, gates and ejector pins are added and after the exposed half of the mould is coated with a release agent, a thin shell of 2-3 millimeters thick of a low temperature molten metal is deposited over it. If they are subjected to high temperature, RP models can soften and distort, and so not all metal deposition techniques can be employed. The most widely used for the replication of RP patterns are:

- **Spray metal deposition (Figure 6.1):** This technique is the most common metal deposition technique and can be divided in two main types: Gas Metal Spraying and Arc Metal Spraying. The former involves a low melting point alloy that passes through a nozzle similar to a paint sprayer. A metal wire, usually lead/tin, is melted by a conical jet of burning gas, atomised and propelled onto the substrate. The second method, also known as the Tafa process, involves a gun in which an electric arc between two wires causes them to melt. The molten material (aluminium or zinc) is then atomised by a compressed gas that sprays it [Dickens, 1996]. Spray metal tools can be used to mould up to 2000 parts in the exact production material. The tools are inexpensive, fast to produce, accurate and capable of handling abrasive materials [3D Systems, 1995].

- **Nickel electroforming:** For this process [Plavcan, 1995; Maley, 1994], the model must first be sprayed with an electrically conductive paint and placed into an acid bath that contains bags of nickel powder. A voltage is applied to the acid bath and nickel is attracted to the conductive paint by electrolysis. This method can be used for a large number of applications to produce injection mould cavities and cores, embossing plates, fine printing plates, seamless belts and laser pumping cavities [3D Systems, 1995].

- **Nickel Vapour Deposition (NVD):** This process is based on the growth of a metal from gaseous vapour. The pattern is heated at temperatures between 110 and 190°C and a nickel carbonyl gas $Ni(CO)_4$ is passed over it. Consequently, a layer of pure nickel is deposited on the pattern and an exact duplicate of its surface is produced. Deposition rates between 0.005 and 0.8 mm/h can be obtained so that a shell mould can be produced rapidly, regardless of its size or complexity [Davy, 1996].

Once a metallic shell has been created using one of the above methods, water cooling lines can be added and the shell is backfilled with epoxy resin or ceramic to improve the strength of the mould. These materials are selected because their coefficient of thermal expansion is close to that of the nickel or zinc the shells are made from. Aluminium powder is usually mixed with the epoxy resin or the ceramic to increase their thermal conductivity. After the backfilling material is cured, it is machined flat. The second half of the tool is built following the same procedure [Jacobs, 1996b].

The main disadvantage of metal spraying is that it is not suitable where the part possesses features such as projections which partially block the metal spray configuration or where recesses may be too deep to spray into completely [Bettany and Cobb, 1995]. For this reason, this process is restricted to models with large and gently curved surfaces. Metal spraying produces economic tooling shells with good reproduction and dimensional qualities but low mechanical strength and high porosity. A way to improve the thermal conductivity of the moulds is to deposit a layer of metal with a higher melting point but better thermal properties over the shell.

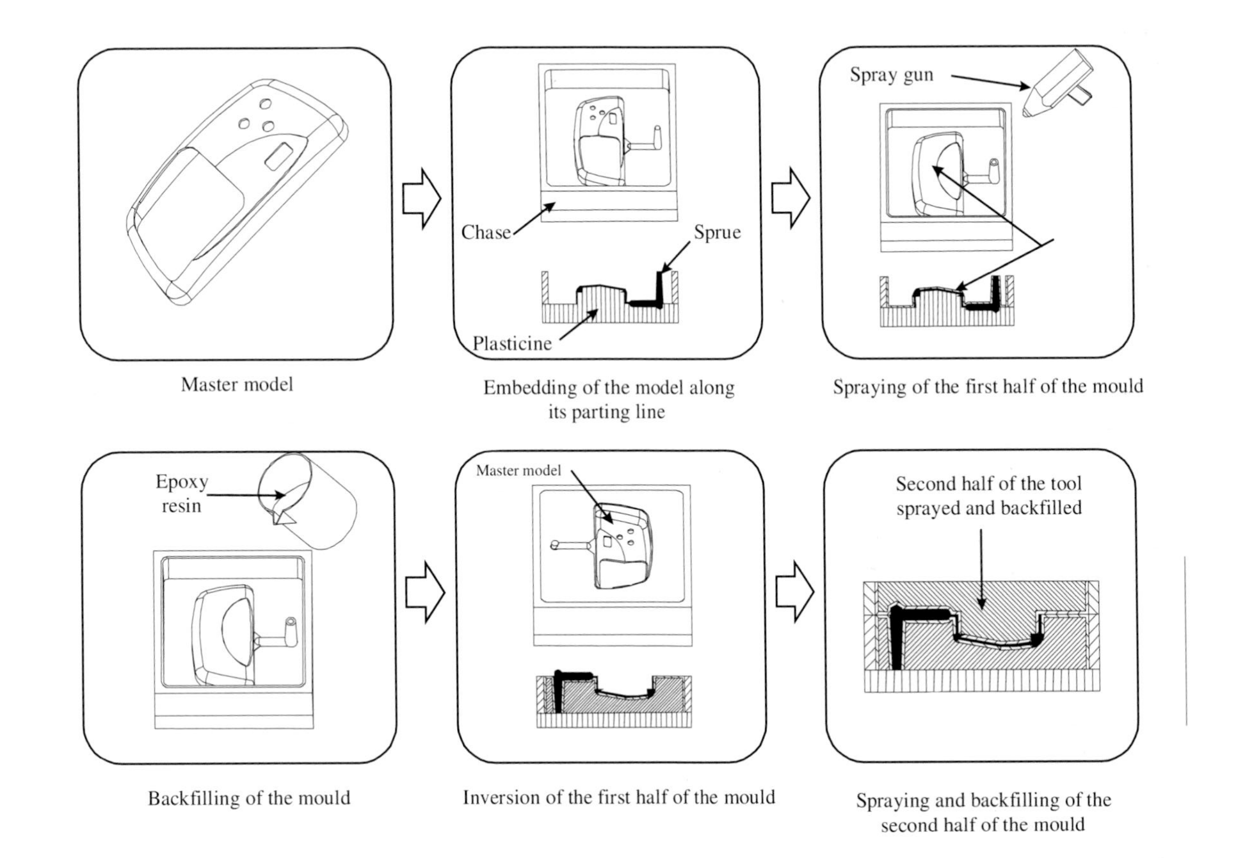

Figure 6.1 Spray Metal Deposition Mould

Nickel has better mechanical and thermal properties than the alloys used for metal spraying but the nickel electroforming deposition rate is slow (about 10μm/h) [Whitward, 1996] and the deposition thickness is dependent on the surface geometry. Areas of the model with deep corners, sharp edges and narrow openings are extremely difficult to plate and can be the areas where the nickel shell will begin to wear [Plavcan, 1995]. NVD does not suffer from the same problems as nickel electroforming but requires models capable of withstanding temperatures of 110°C.

Metal deposition is one of the more mature and effective tooling techniques presented in this paper. Tools made this way allow production quantities of several thousand parts and can be found in many applications including sheet metal forming, injection moulding, blow moulding and compression moulding.

6.3 RTV Tools

RTV tools are an easy, relatively inexpensive and fast way to fabricate prototype or pre-production tools. RTV tools are also known as silicone rubber moulds. The fabrication of RTV moulds usually includes the following main steps [Jacobs, 1996]:

1. Producing a pattern. Any RP method can be employed.
2. Adding venting and gating to the pattern.
3. Setting-up the pattern in a mould box with a parting line provided in a plasticine.
4. Pouring silicone rubber to form one half of the mould.
5. Inverting the first half of the mould and removing the plasticine.
6. Pouring silicone rubber to produce the second half of the mould.

There are two types of silicone used in this process: tin- and platinum-based silicones. Tin-based silicone is generally less expensive and more durable.

RTV tools can be utilised for moulding parts in wax, polyurethane and a few epoxy materials. The process is best suited for projects where form, fit, or functional testing can be done with a material which mimics the characteristics of the production material.

Another form of RTV moulding known as Vacuum Casting is widely used for producing accurate silicone tools for casting parts with fine details and very thin walls. The process requires initial investment in a vacuum chamber with two sections. The upper section is for mixing the resin and the lower is for casting the resin into the mould. A vacuum chamber manufactured by MCP Systems is shown in Figure 6.2.

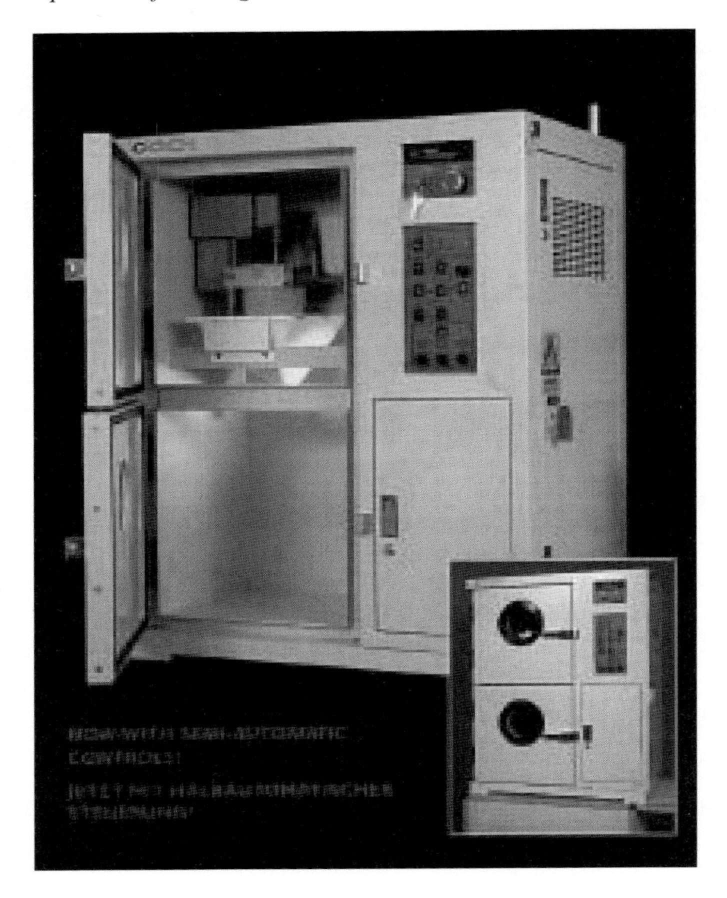

Figure 6.2 MCP vacuum casting chamber (Courtesy of MCP)

The MCP vacuum casting process includes nine steps as shown in Figure 6.3 [3D Systems, 1995]:

1. The first step is to produce a pattern using any of the available RP processes (SLA, SLS, FDM, etc.).
2. The pattern is fitted with a casting gate and set up on the parting line, and then suspended in a mould casting frame.
3. Once the two-part silicone-rubber is de-aerated and then mixed, it is poured into the mould casting frame around the pattern.
4. The mould is cured inside a heating chamber.

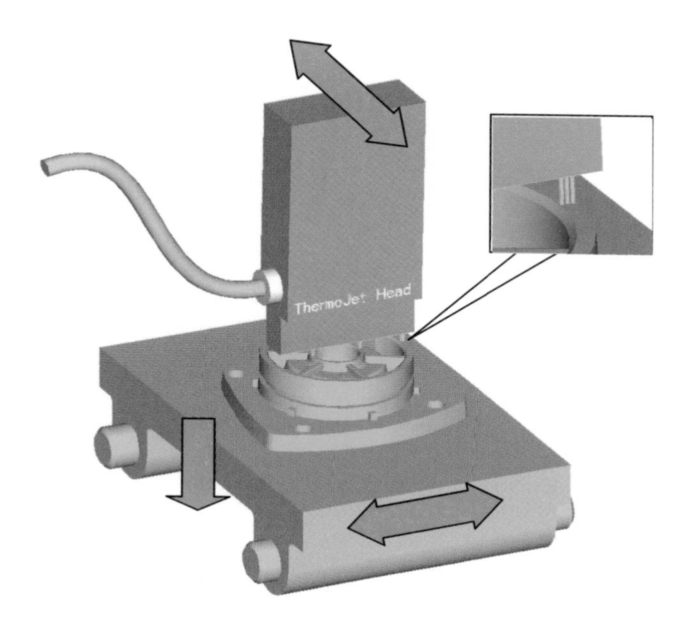

Figure 2.8 Multi Jet Modelling Head, p.29

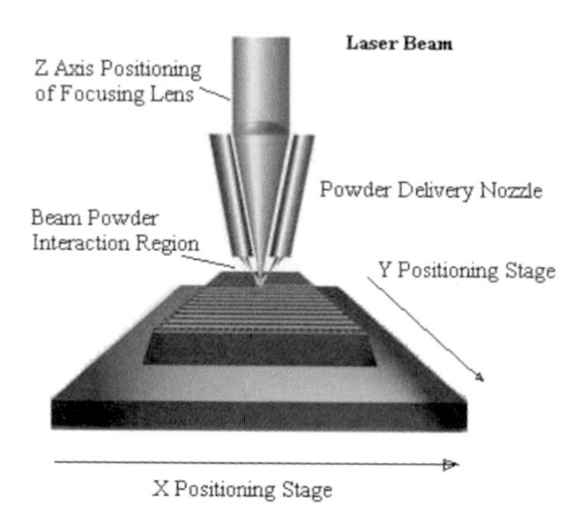

Figure 2.12 LENS™ process (Courtesy of Optomec Design Co), p.35

Figure 3.25 LENS^{TM} process (Courtesy of Optomec Design Co), p.65

Figure 3.28 Parts produced employing LENS and stereolithography (Courtesy of Optomec Design Co), p.67

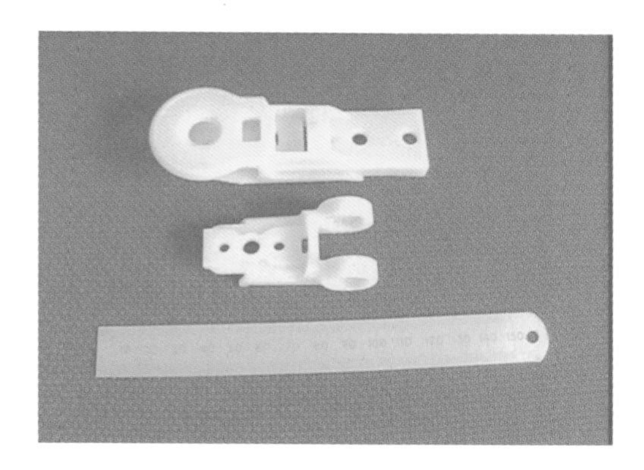

Figure 4.3 Example of model built using the ThermoJet printer, p.73

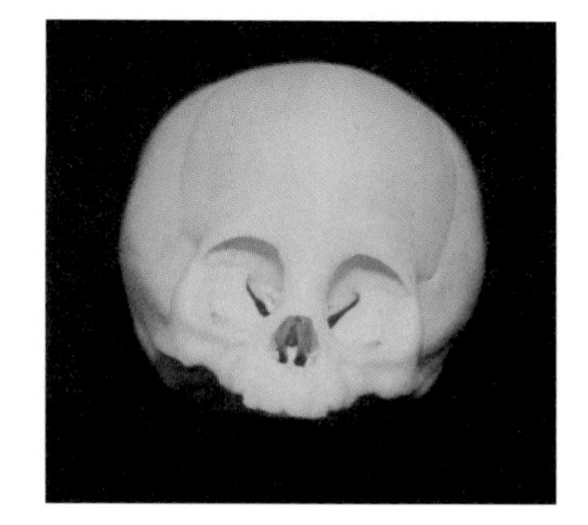

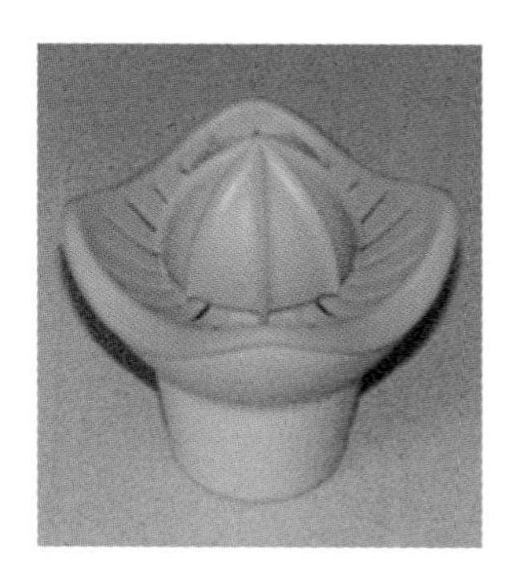

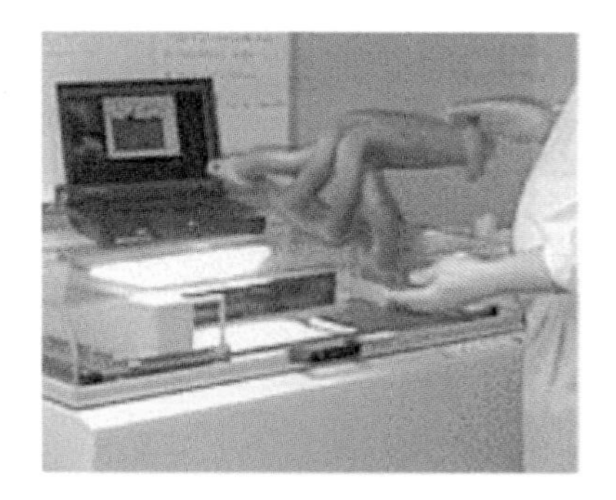

Figure 4.10 Examples of parts built using the 3DP process
(Courtesy of Z Corp.), p.77

Figure 4.7 A ring produced employing PatternMaster (Courtesy of Sanders Prototype, Inc.), p.75

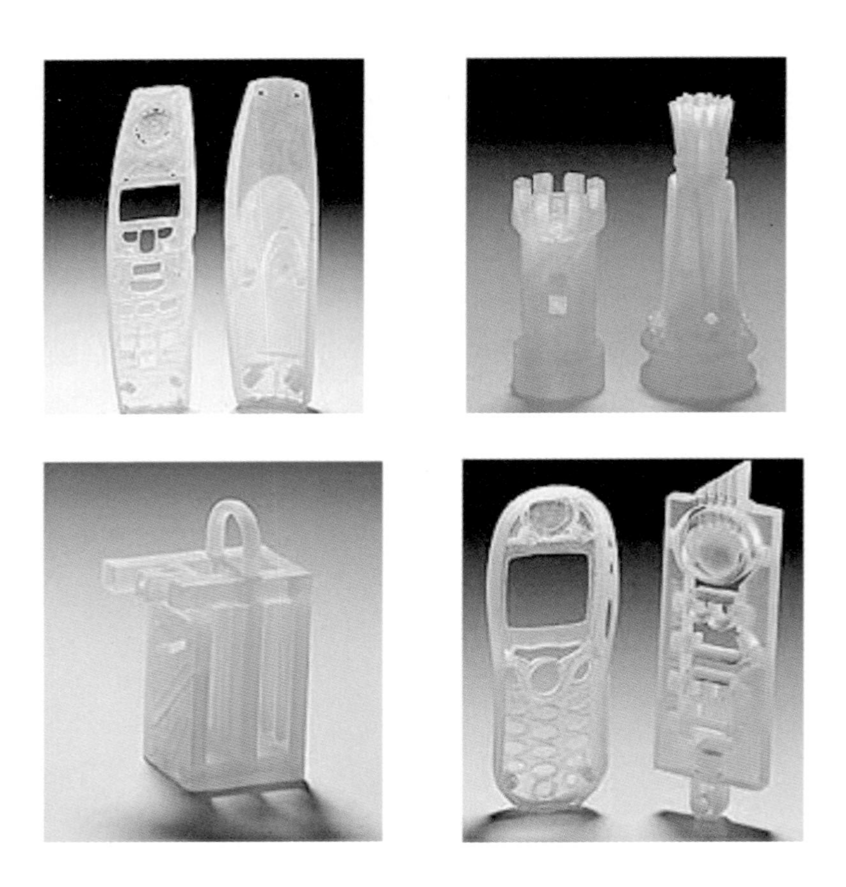

Figure 4.17 Example of models produced using the Objet Quadra (Courtesy of Objet Geometry), p.84

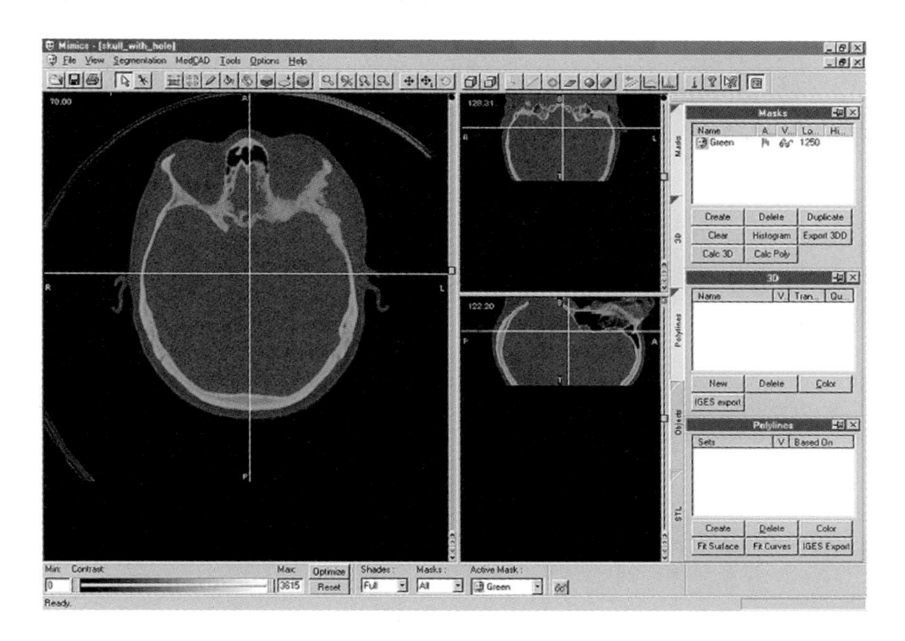

Figure 5.17 User interface in the MIMICS software (Courtesy of Materialise), p.99

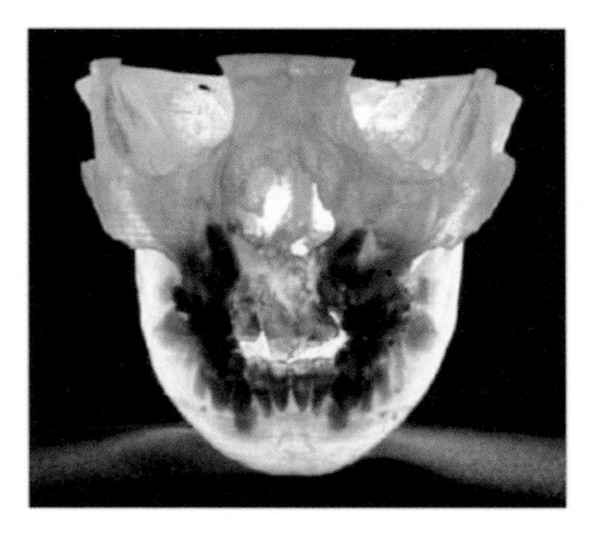

Figure 5.18 A medical model built using the Stereocol resin (Courtesy of Materialise), p.99

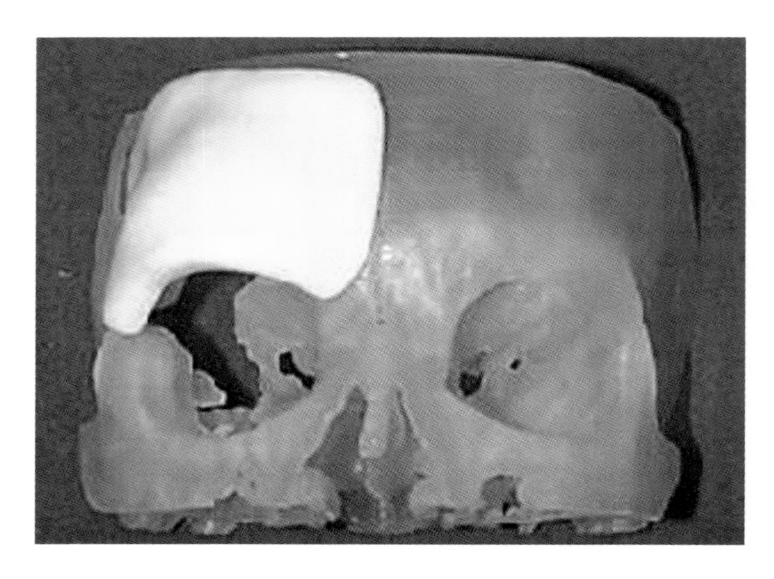

Figure 5.19 The SLA model with the resection template (Courtesy of the British Journal of Neurosurgery), p.100

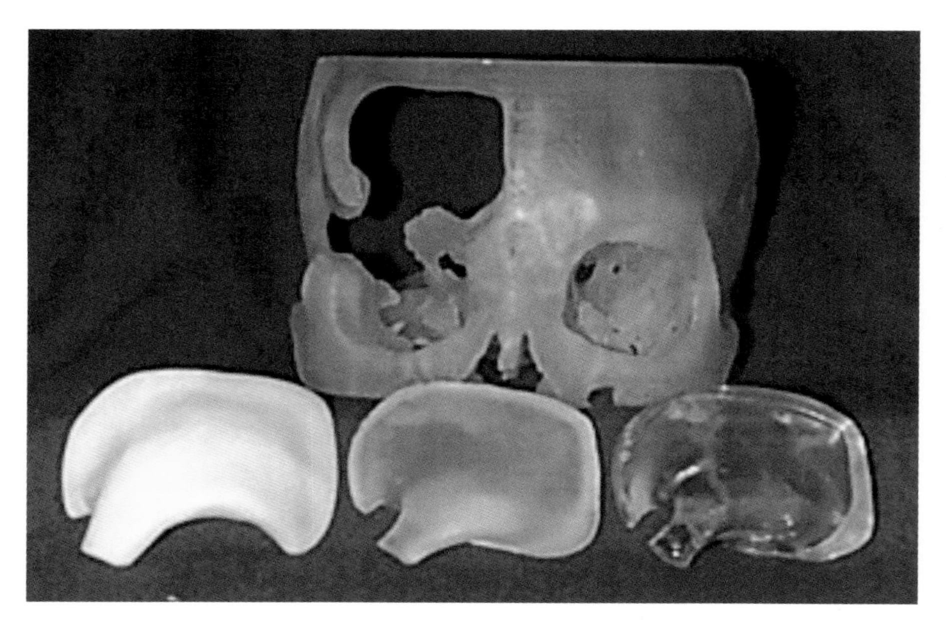

Figure 5.20 The SLA model together with the template and the implant (Courtesy of the British Journal of Neurosurgery), p.101

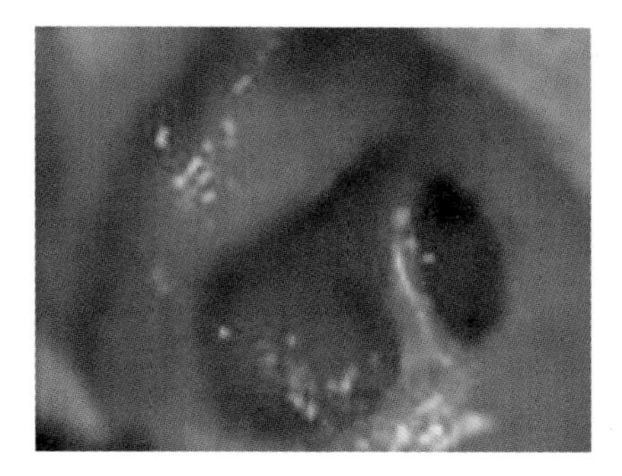

A: A hole resulted from the irradiation of a tumor in the mouth cavity. This hole has to be filled by an implant to allow the patient to breathe and eat normally.

B: The soft tissue surrounding the cavity was modelled by CT scanning. This model was used to design a tool for direct moulding of the implant.

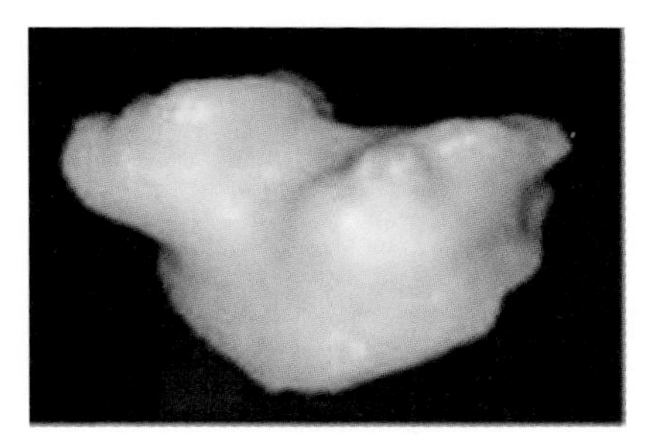

C: A silicon implant was moulded from the tool.

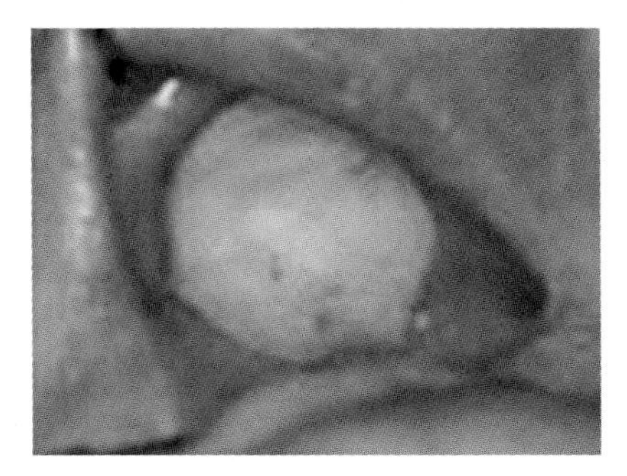

D: Without surgery, the deformable silicon prosthesis was implanted. Magnets were used to fix the prosthesis to a hard dental implant.

Figure 5.21 Fabrication of obturator prosthesis using RP techniques. Case study presented by Dr L.L. Visch from Daniel den Hoed Kliniek Rotterdam (Courtesy of Materialise), p.102

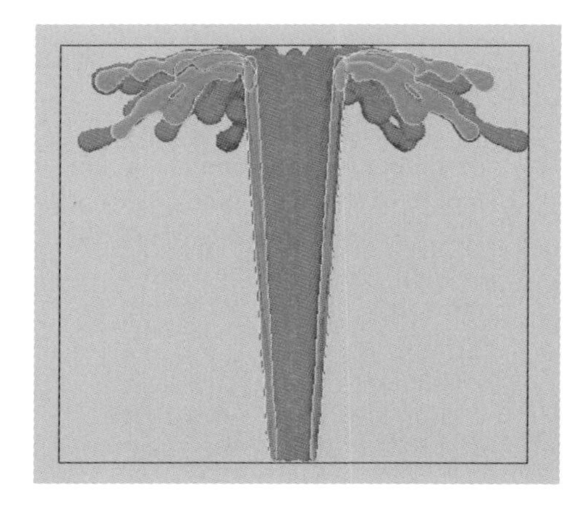

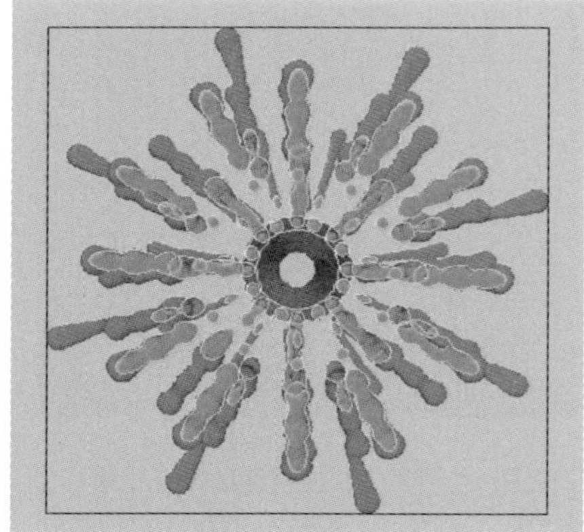

Figure 5.22 Cross-sections of the 3D model of a water splash (Courtesy of the CALM project), p.103

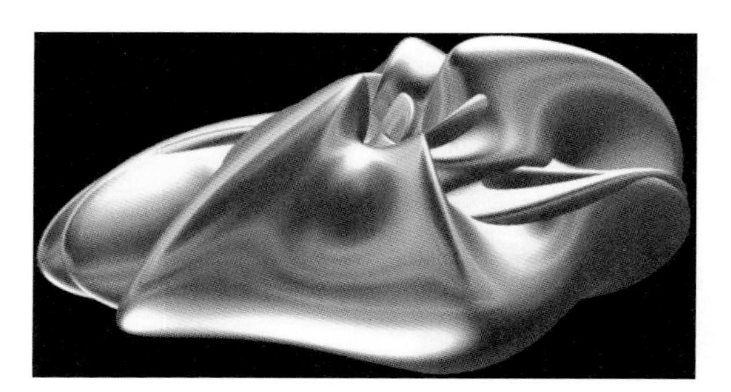

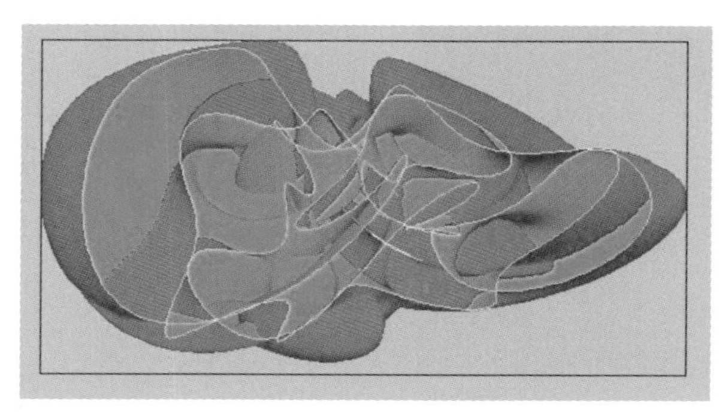

Figure 5.24 3D shaded image and a cross-section of the cybersculpture (Courtesy of the CALM project), p.104

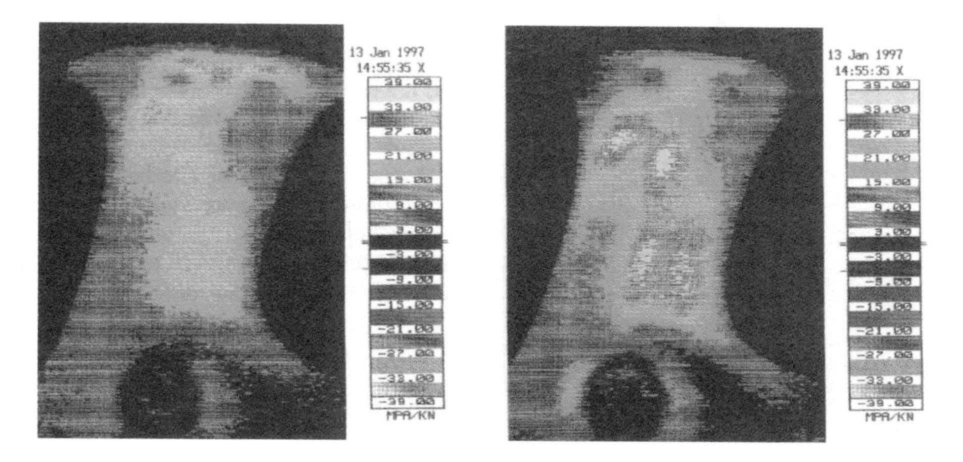

Figure 5.28 Thermographic plots of aluminium rim (left) and SLS glass-filled nylon rim (right) [Gartzen et al., 1998], p.107

Figure 5.29 The frozen stress distribution for a model of an aeroengine turbine rotor [3D Systems, 1994], p.108

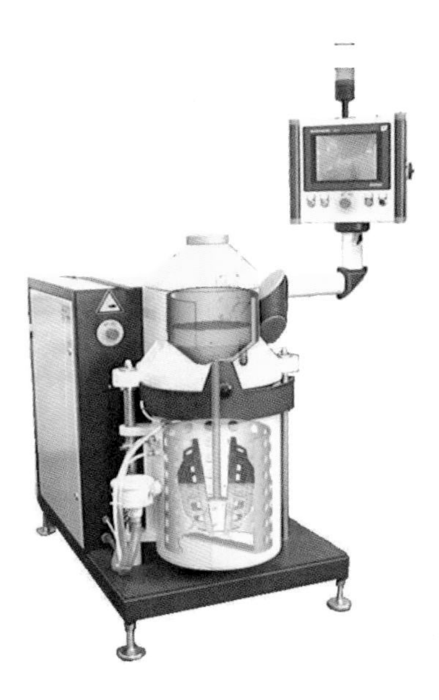

Figure 6.9 MCP Metal Part Casting System, p.127

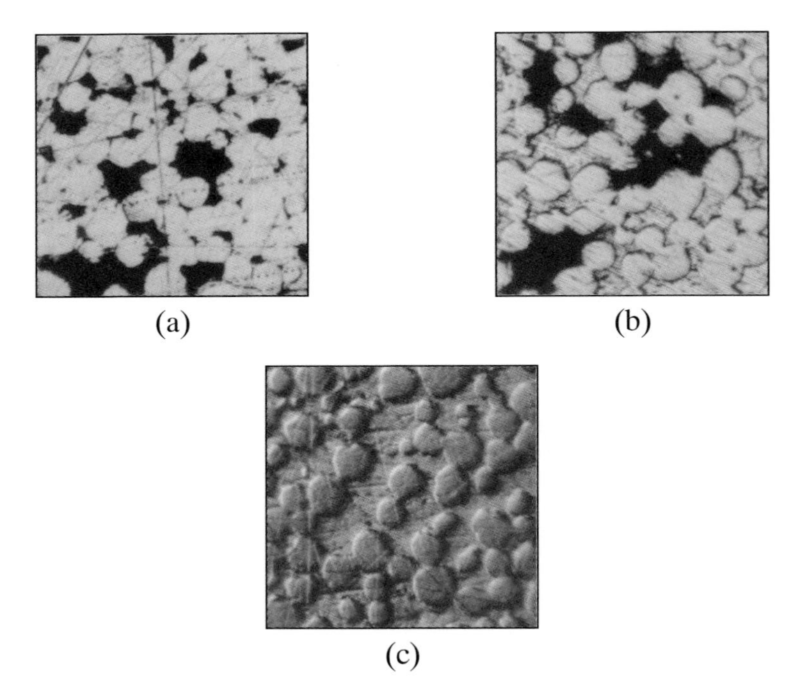

Figure 7.4 Optical micrograph of a RapidSteel part surface
(a - Sintered Part, b - Partially copper infiltrated, c- Fully copper infiltrated), p.142

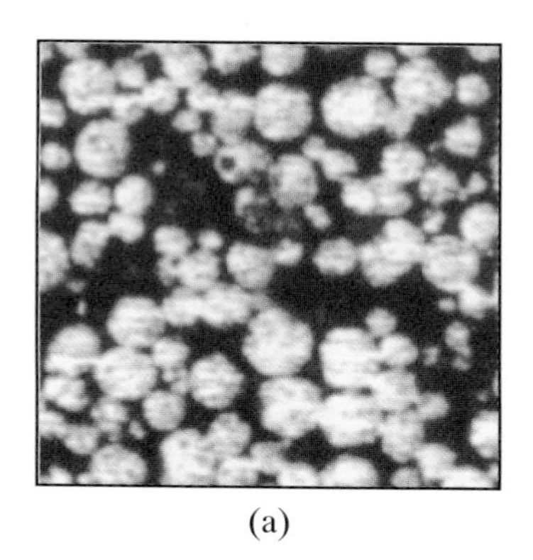

(a)

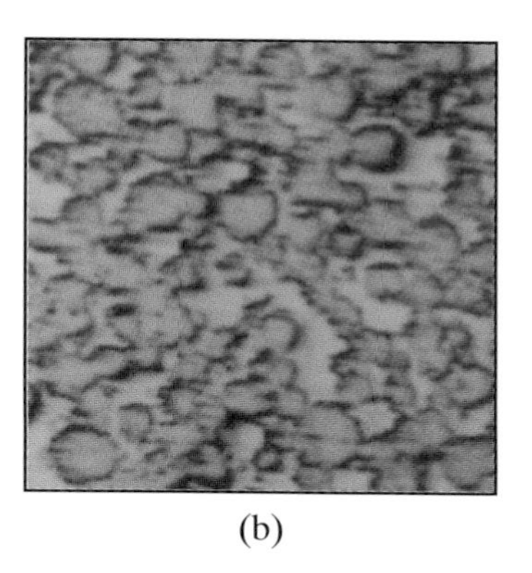

(b)

Figure 7.7 Optical micrograph of a RapidSteel 2.0 part surface (a - Sintered part, b - Fully bronze infiltrated part), p.145

Figure 7.14 SandForm™ core (Courtesy of DTM), p.151

1. CAD model of a casting tree

2. CAD model of a ceramic mould

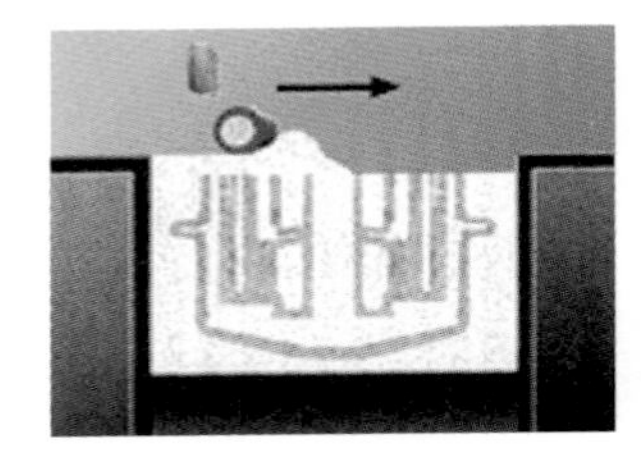

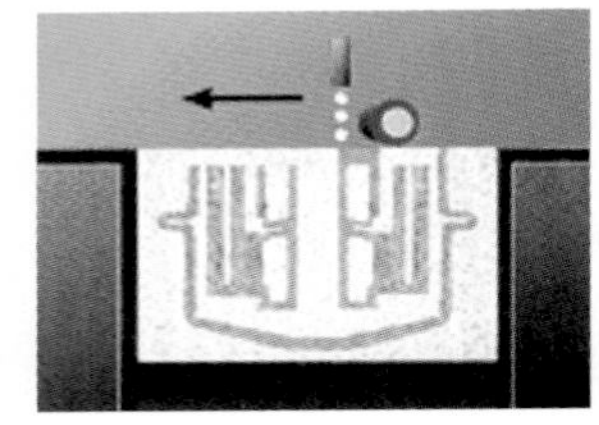

3. Building the mould using 3D printing process

4. Removal of the unbound powder

5. Casting the mould

Figure 7.19 Direct Shell Production Casting (Courtesy of Soligen), p.156

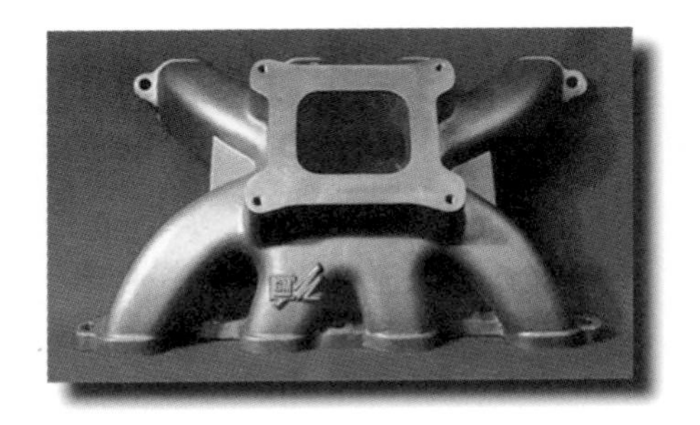

Figure 7.20 Examples of ceramic moulds and castings fabricated using DSPC (Courtesy of Soligen), p.157

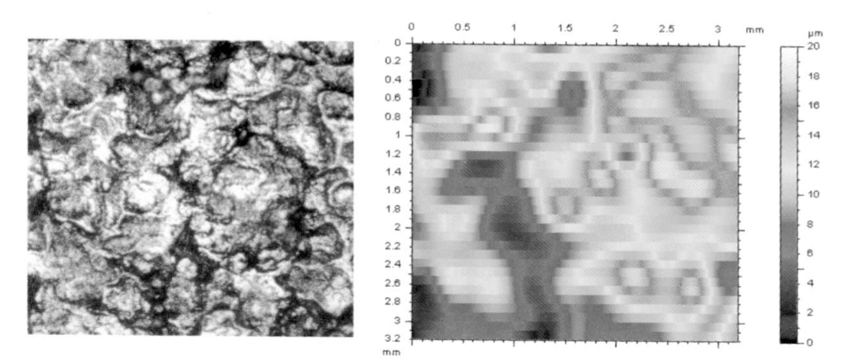

Figure 8.6 Micrograph and waviness colour coded map of the square feature surface of the EOSINT M insert after 16000 injections (magnification 1000x), p.169

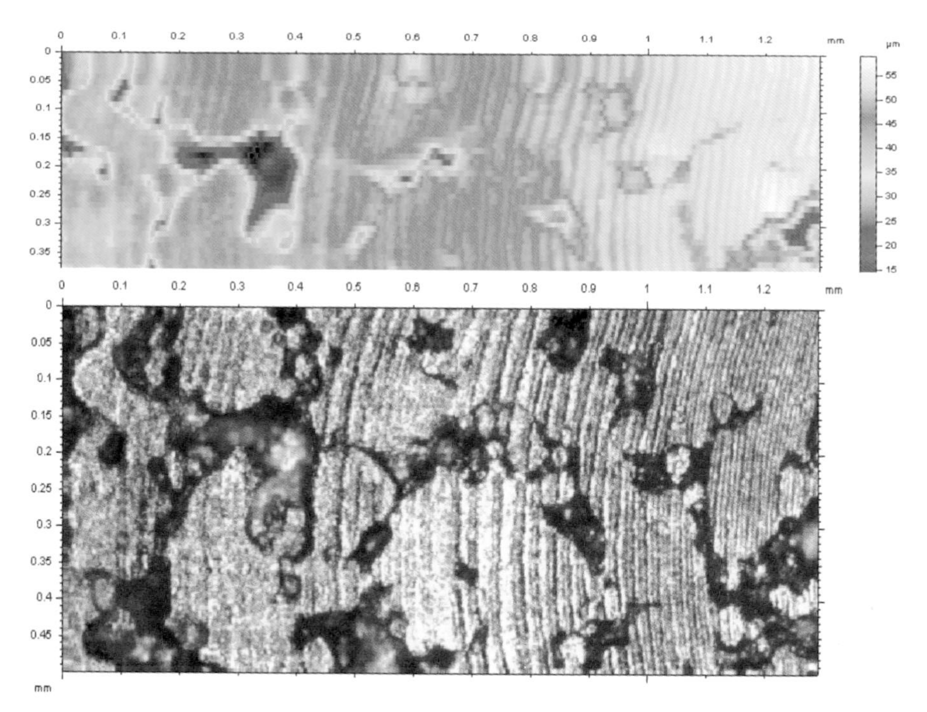

Figure 8.8 Colour coded depth map and micrograph (magnification 1000x) of the gate of the non-coated EOSintM insert after 16000 injections, p.171

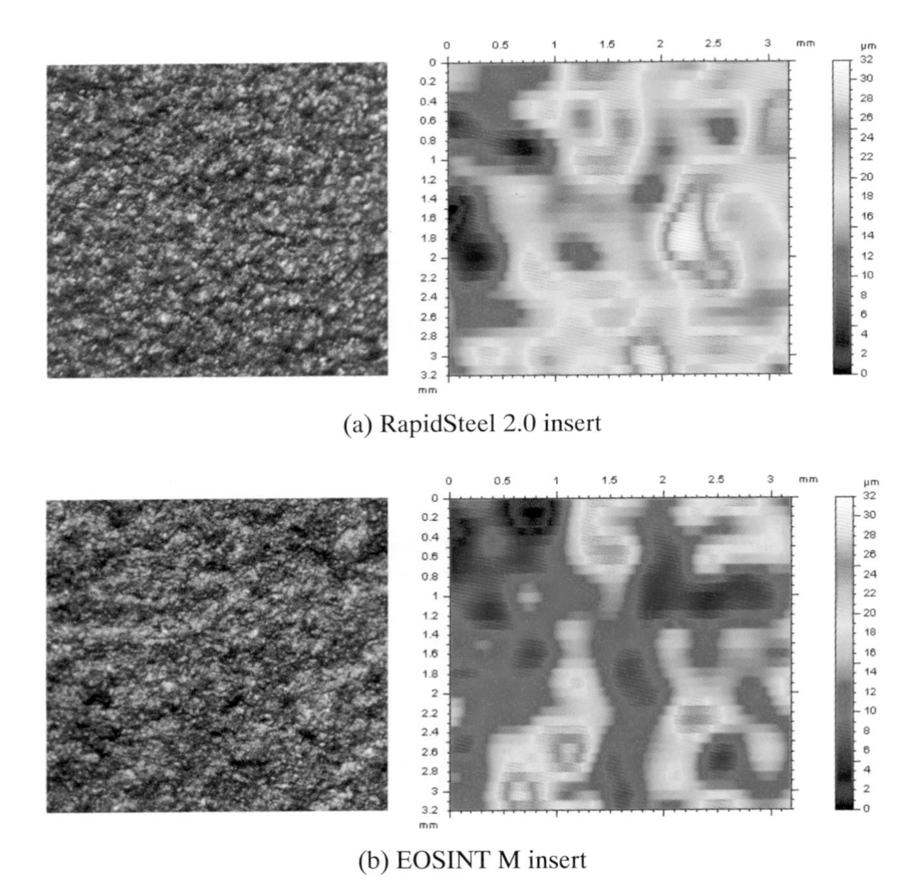

(a) RapidSteel 2.0 insert

(b) EOSINT M insert

Figure 8.10 Surface (magnification 250x) and waviness of the square feature for the Molybdenum-coated RapidSteel 2.0 and EOSINT M inserts after 8000 injections, p.172

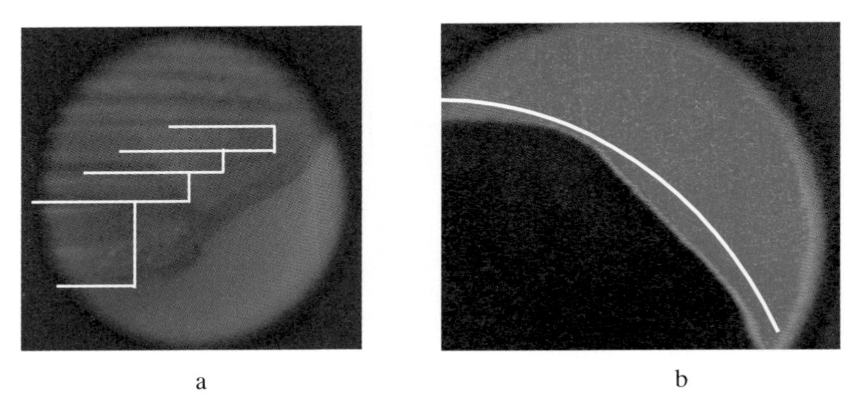

a b

Figure 9.7 Over-curing effects on accuracy (a - thicker bottom layer, b - deformed hole boundary), p.192

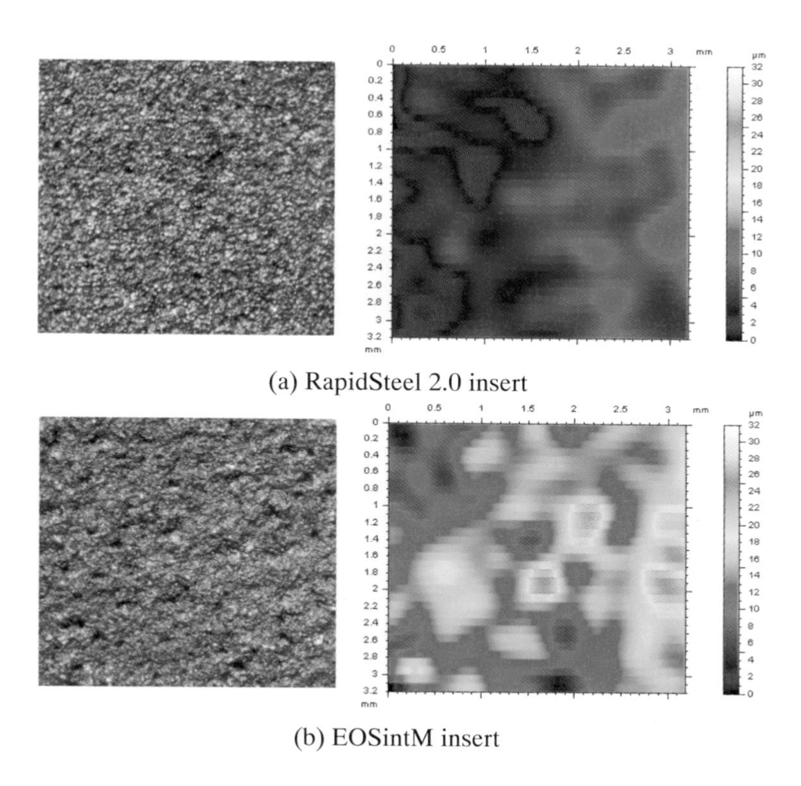

Figure 8.11 Surface (magnification 250x) and waviness of the square feature for the Stellite-coated RapidSteel 2.0 and EOSINT M inserts after 8000 injections, p.173

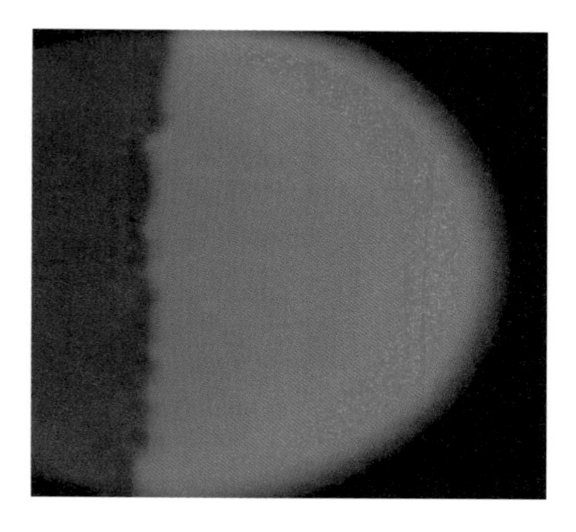

Figure 9.9 Vertical wall, p.193

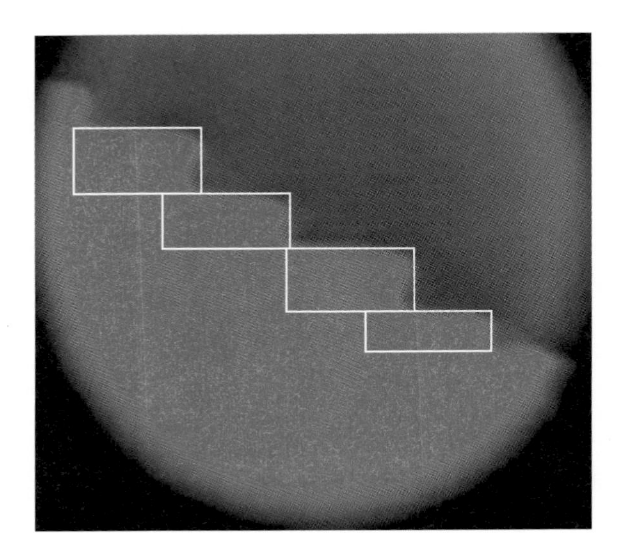

Figure 9.10 Effects on curves, p.193

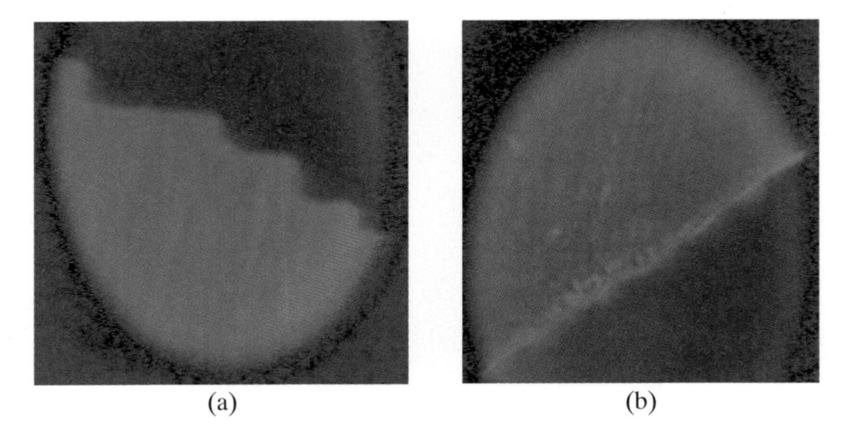

(a) (b)

Figure 9.16 Stair-steps around the periphery (a - facing upwards; b - facing downwards), p.198

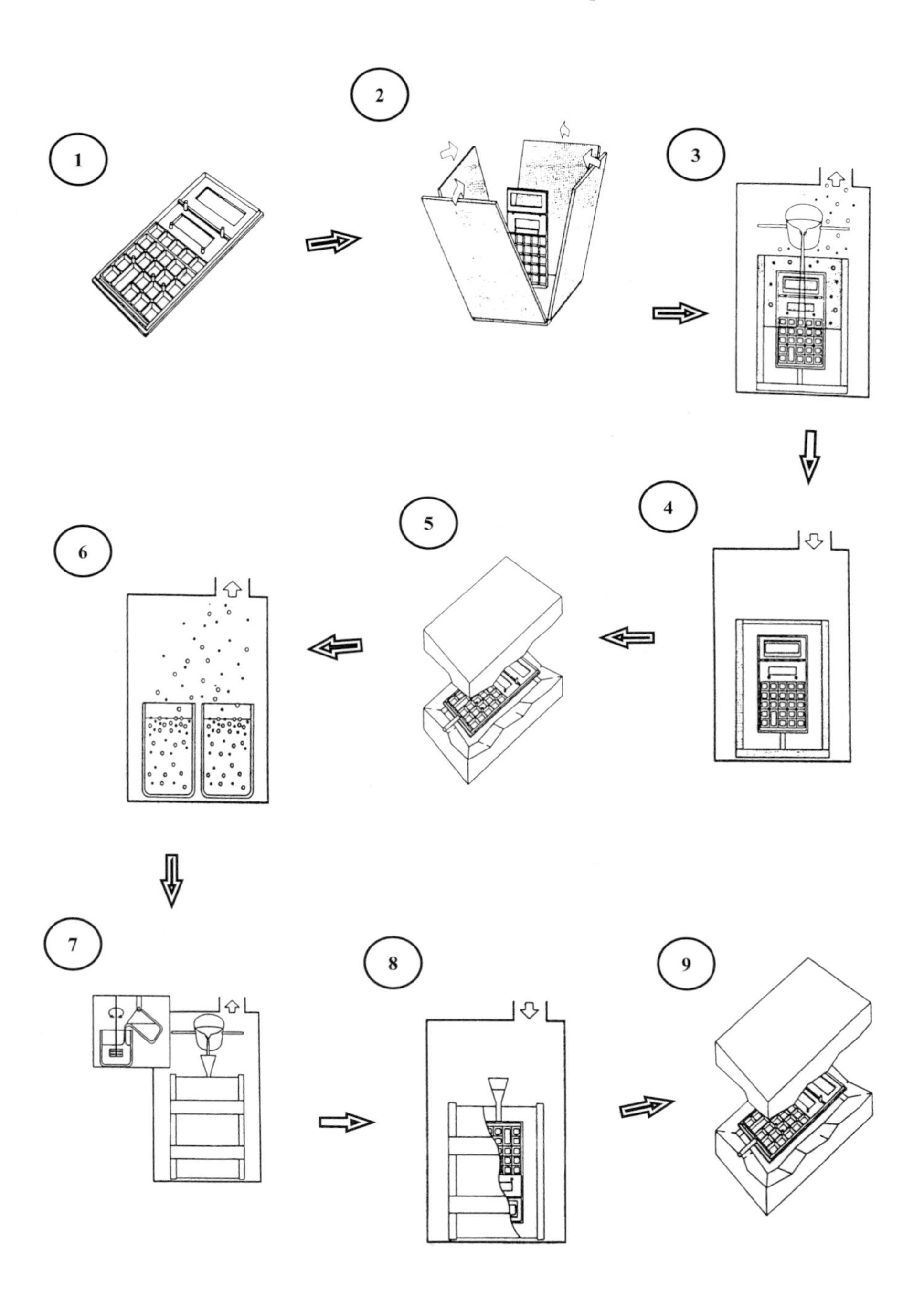

Figure 6.3 MCP vacuum casting process (Courtesy of MCP)

5. The pattern is removed from the silicone mould by cutting along the parting line.

6. The urethane resin is measured, dye is added for coloured components and casting funnels placed. Then, the mould is closed and sealed.

7. The computer-controlled equipment mixes and pours the resin inside the vacuum chamber. Because this takes place in a vacuum, the mould is filled completely without leaving any airpockets or voids.

8. After casting the resin the mould is moved to the heating chamber for 2 to 4 hours to cure the urethane part.

9. After hardening, the casting is removed from the silicone mould. The gate and risers are cut off to make an exact copy of the pattern. If required the component can be painted or plated.

Vacuum castings are precise replicas of the patterns, dimensionally accurate without blemishes, with all profiles and textures faithfully reproduced. A variety of resins specially-formulated for vacuum casting are available on the market to offer various characteristics in hardness, toughness, flexibility and temperature resistance.

6.4 Epoxy Tools

Epoxy tools are used to manufacture prototype parts or limited runs of production parts. Epoxy tools are used for [3D Systems, 1995]:

- Moulds for prototype injection plastics;
- Moulds for castings;
- Compression moulds;
- Reaction injection moulds.

The fabrication of the mould begins with the construction of a simple frame around the parting line of the RP model (Figure 6.4). Sprue gates and runners can be added or cut later on, once the mould is finished. The exposed surface of the model is coated with a release agent and epoxy is poured over the model. Aluminium powder is usually added to the epoxy resin and copper hose cooling lines can also be placed

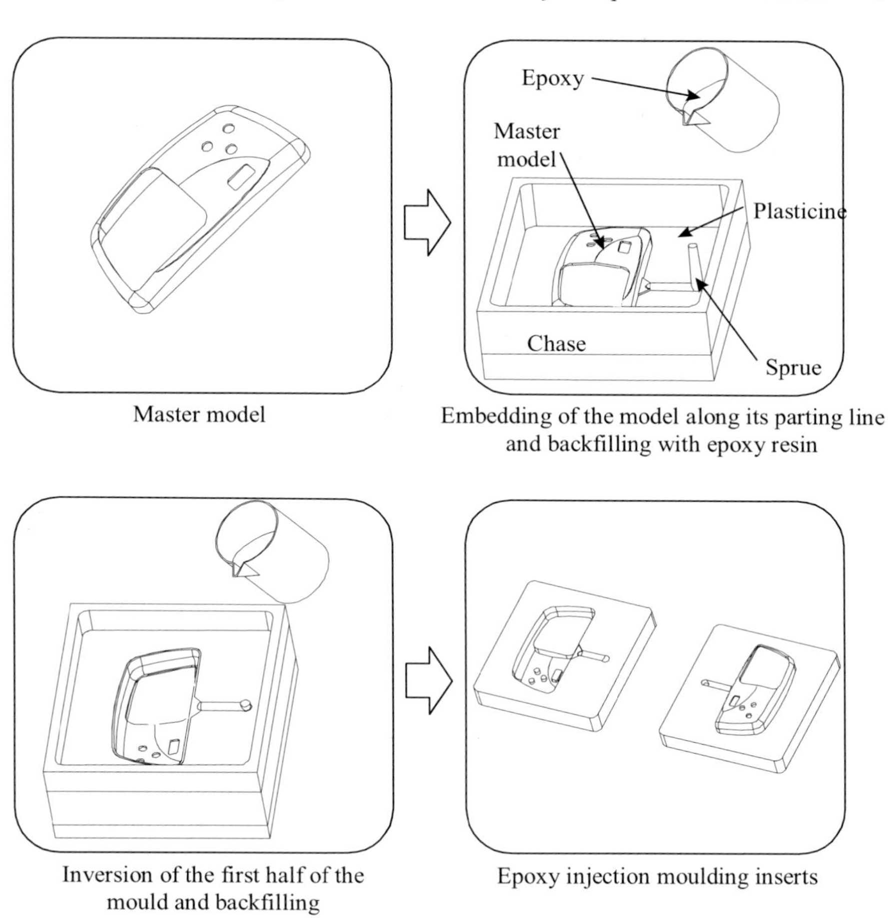

Master model

Embedding of the model along its parting line and backfilling with epoxy resin

Inversion of the first half of the mould and backfilling

Epoxy injection moulding inserts

Figure 6.4 Epoxy Mould

at this stage to increase the thermal conductivity of the mould. Once the epoxy has cured, the assembly is inverted and the parting line block is removed, leaving the pattern embedded in the side of the tool just cast. Another frame is constructed and epoxy poured to form the other side of the tool. When the second side of the tool is cured, the two halves of the tool are separated and the pattern is removed [Mueller, 1995].

Another approach, known as Soft-Surface® rapid tool (Figure 6.5) involves machining an oversized cavity in an aluminium plate. This offset allows for the introduction of the casting material, which may be poured into the cavity after suspending the model in its desired position and orientation [Comeau and Dobson, 1996]. Some machining is required for this method and this can increase the mould building time but the advantage is that the thermal conductivity is better than for all-epoxy moulds.

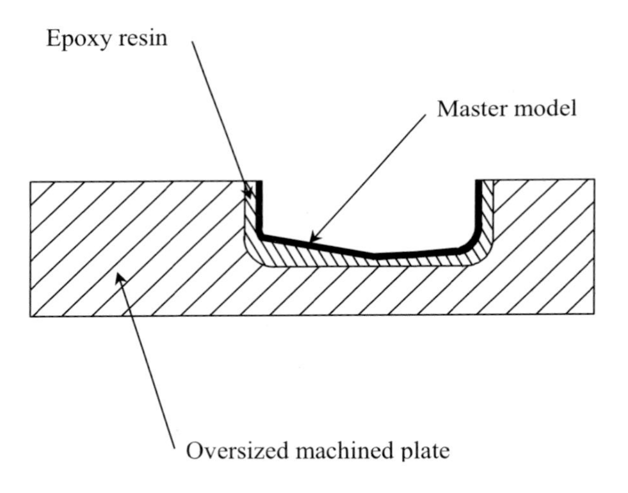

Figure 6.5 Soft-Surface®

Unfortunately, epoxy curing is an exothermic reaction and it is not always possible directly to cast epoxy around an RP model without damaging it. In this case, a silicone RTV mould is cast from the RP pattern and a silicone RTV model is made from the mould and is used as the pattern for the aluminium filled epoxy tool. A loss of accuracy occurs during this succession of reproduction steps. An alternative process is to build an RP mould as a master so that only a single silicone RTV reproduction step is needed [Kamphuis and VanHeil, 1996]. Because epoxy tooling requires no special skill or equipment, it is one of the cheapest techniques available.

It is also one of the quickest. Several hundred parts can be moulded in almost any common casting plastic material.

Epoxy tools have the following limitations [3D Systems, 1995]:

- Limited tool life;
- Poor thermal transfer;
- Tolerance dependent on master patterns;
- Aluminum-filled epoxy has low tensile strength.

The life of injection plastic aluminium-epoxy tools for different thermoplastic materials is given in Figure 6.6 [3D Systems, 1995].

Material	**Tool Life (shots)**
ABS	200-3000
Acetal	100-1000
Nylon	250-3000
Nylon (glass filled)	50-200
PBT	100-500
PC/ABS Blends	100-1000
Polycarbonate	100-1000
Polyethylene	500-5000
Polypropylene	500-5000
Polystyrene	500-5000

Figure 6.6 Approximate aluminium-epoxy tool life [3D Systems, 1995]

6.5 Ceramic Tools

Instead of epoxy, any plaster ceramics can also be cast around a master to produce a tool cavity. Ceramic tools can be employed in plastics processing, metal forming and metal casting [Dickens, 1996]. In making ceramic tools, the amount of water used has to be controlled to avoid excessive shrinkage as the material sets. Recently, attention has been focused on non-shrinking ceramics. These Calcium Silicate-based Castable (CBC) ceramics were initially developed for applications where metal spraying was not suitable.

The stages of producing the two halves of a CBC ceramic tool differ slightly from the epoxy mould procedure described earlier. CBC ceramics only generate a small amount of heat during curing (approximately 50°C). This allows them to be poured directly over the RP master without damaging it. The two halves of the mould must be vacuum cast to avoid air bubbles and a vibration table can help to pack the material. After about one hour, the RP pattern can be removed and the ceramic tool is cured for about 24 hours in an oven. Once the ceramic is fully cured, the back surfaces of the two mould halves are machined flat and guides are drilled to receive the ejector pins [Jacobs, 1996b].

Ceramics are porous materials, which is not desirable when the tool is used to mould very adhesive polymers. Various surface treatments can be carried out to reduce the porosity including the application of a dry film lubricant, a release agent, silicone, or PTFE [Bettany and Cobb, 1995]. Ceramic tools are brittle and must be handled with care. Finely chopped fibres are often added to enhance fracture toughness and tensile strength, as well as aluminium fillers to increase the thermal conductivity. In this way, a tool can be used to produce several hundred parts and injection moulding cycle times as low as 30 seconds can be achieved.

The main advantage of this process, apart from the low cost of the ceramics used, is the short time needed to build a mould. Some beta-tested CBC ceramics have been reported with a curing time of a few hours [Bettany and Cobb, 1995], which could enable an injection tool to be made in one day after obtaining the RP model.

6.6 Cast Metal Tools

Metal moulds are generally time-consuming and expensive to machine, but by combining RP techniques with casting techniques, some zinc or aluminium alloy moulds can be rapidly made.

- **Investment casting:** The use of RP sacrificial models for investment casting was one of the first applications of RP. Nowadays, models for investment casting can be made on almost every RP machine. They can be obtained directly without any change to the building process (LOM), by modifying the building style (Quickcast™), or by using a special material (SLS, FDM). Another technique is to build the ceramic shell that will be used for investment casting (3-D Printing).

- **Die-casting:** Proceeding in the same way as for ceramic tooling, it is possible to fabricate a ceramic mould to cast a metal alloy mould. The ceramic mould then behaves like a die cast mould but can be used for the fabrication of one metal mould only.

- **Spin-casting:** The spin-casting process consists of injecting a material through a central sprue into a mould that is rotated at high speed (Figure 6.7). Spin-casting moulds for metal parts are made of heat vulcanised silicone. The heat that is given out during the fabrication of such moulds is too high for usual RP patterns. For this reason, the fabrication of a metal part using spin-casting consists of several steps. First, an RTV rubber mould is made from the RP master. From this mould, a tin based metal alloy part is cast and is used as a model for the fabrication of a heat-vulcanised silicone mould [Schaer, 1995]. This final mould can produce spin-cast zinc alloys parts that have similar physical strength properties to both die cast aluminium parts and die cast Zamak zinc parts [Mosemiller and Schaer, 1997].

The aim when using an RP model for metal casting is to make a mould as similar as possible to the final mould so that only finishing is required. In this way, time and machining are saved compared with traditional mould making methods.

Unlike the tooling methods presented previously, these metal tools have relatively good strength and thermal conductivity. This allows normal clamping forces and injection pressures during the moulding cycle. The injection moulding conditions can then be considered similar to production conditions but the life of the mould is usually limited to below a thousand parts [Erickson, 1996].

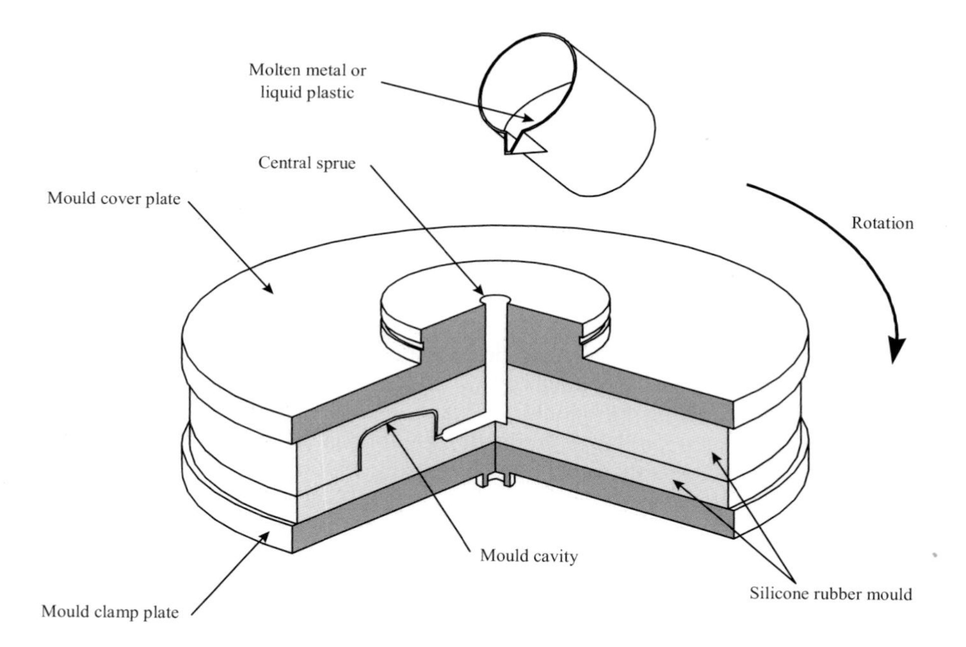

Figure 6.7 Spin-casting

6.7 Investment Casting

The investment casting process is used to cast complex and accurate parts. The process was invented by the early Egyptians and was called "lost wax". As the name indicates, wax patterns are used to define the part shape that are then melted away. It is also possible for patterns to be produced from foam, paper, polycarbonate and other RP materials which can be easily melted or vaporised. The investment casting process includes the following main steps as shown in Figure 6.8:

1. Multiple patterns are produced.

2. The patterns are assembled as a group on a "tree" where they are gated to a central sprue.

3. The tree of patterns is dipped in a slurry of ceramic compounds to form a coating. Then, refractory grain is sifted onto the coated patterns to form the shell.

4. Step 3 is repeated several times to obtain the desired shell thickness (5-10 mm) and strength.

5. After the tree has set and dried, the patterns are melted away or burned out of the shell, resulting in a cavity.

6. Molten metal is poured into the shell to form the parts.

7. The ceramic shell is broken away to release the castings.

8. Finally, the castings are removed from the sprue and the gate stubs are ground off.

The above process is known as shell investment casting. Another form of investment casting is solid flask investment casting. The latter employs solid flask moulds instead of shells. In addition, the moulds are filled while applying a vacuum differential pressure method. An automated system for solid flask investment casting manufactured by MCP is shown in Figure 6.9. Production of castings employing this system involves the following processing steps [MCP, 2000]:

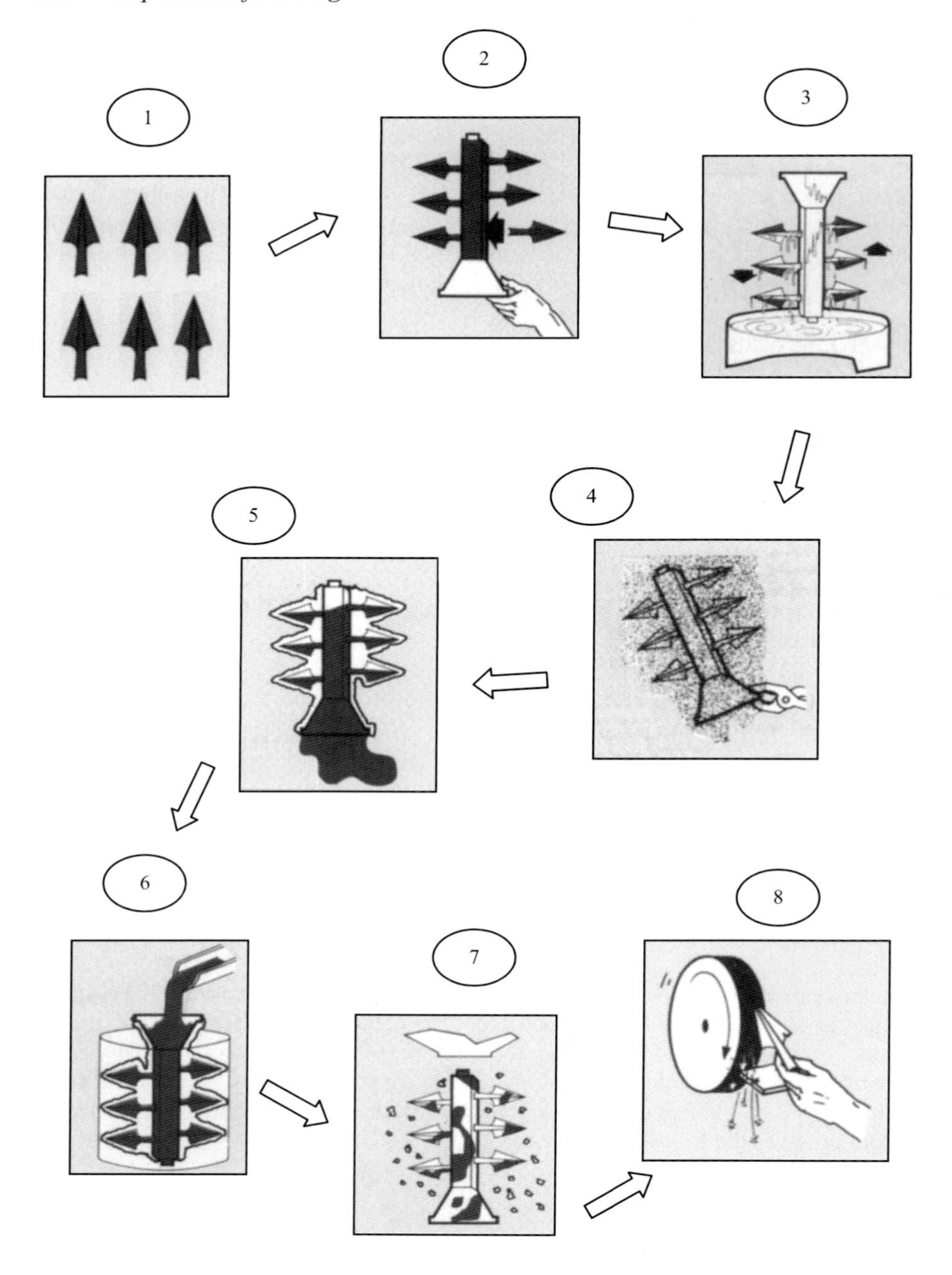

Figure 6.8 Investment casting [Tritech, 1997]

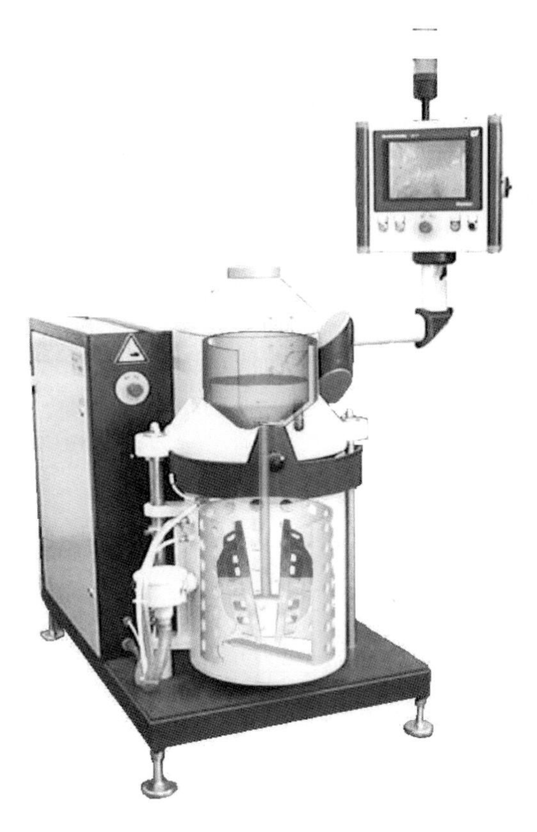

Figure 6.9* MCP Metal Part Casting System

1. A master is built using any of the available RP processes.

2. A silicone mould is produced from the master under vacuum in the MCP Vacuum Casting System (see Section 6.3).

3. Multiple patterns are cast from the silicone mould in the same system under vacuum.

4. The patterns are gated to a central sprue to create a pattern cluster. The cluster is then placed in a flask.

5. Ceramic embedding material is poured around the pattern cluster under a vacuum in a special vacuum chamber to avoid creating bubbles.

6. After the ceramic mould has set and dried, the flask is placed in a furnace to melt out the patterns.

7. The flask is next placed in the casting chamber of the MCP Metal Part Casting system. The melting chamber is designed to provide a melting pressure independent of the casting chamber. Thus, the mould is filled with metal by vacuum differential pressure.

8. The ceramic mould is broken away from the castings using a water jet. Then, the sprue and gate stubs are removed from the castings.

The MCP Metal Part Casting system is manufactured in three unit sizes: MPA 150, MPA 300 and MPA 1000. The technical specifications of these three models are given in Figure 6.10.

Unit Type	Casting Volume of Metal	Flask Size
MPA 150	1.5 litres	Ø 250 x 350 mm
MPA 300	3.0 litres	Ø 350 x 500 mm
MPA 1000	10.0 litres	Ø 550 x 700 mm

Figure 6.10 Technical specifications of the MCP Metal Part Casting systems

6.8 Fusible Metallic Core

Fusible metallic core technology is a new method for forming complex, hollow, one-piece plastic components which may be difficult to produce by any other method. This technology can be considered as a variation of the investment casting process. The difference between the two processes is in the material of the sacrificial patterns employed. In particular, low melting point alloys are used instead of wax. RP techniques can be used to build part and core models that assist in fabricating the casting dies to make the cores.

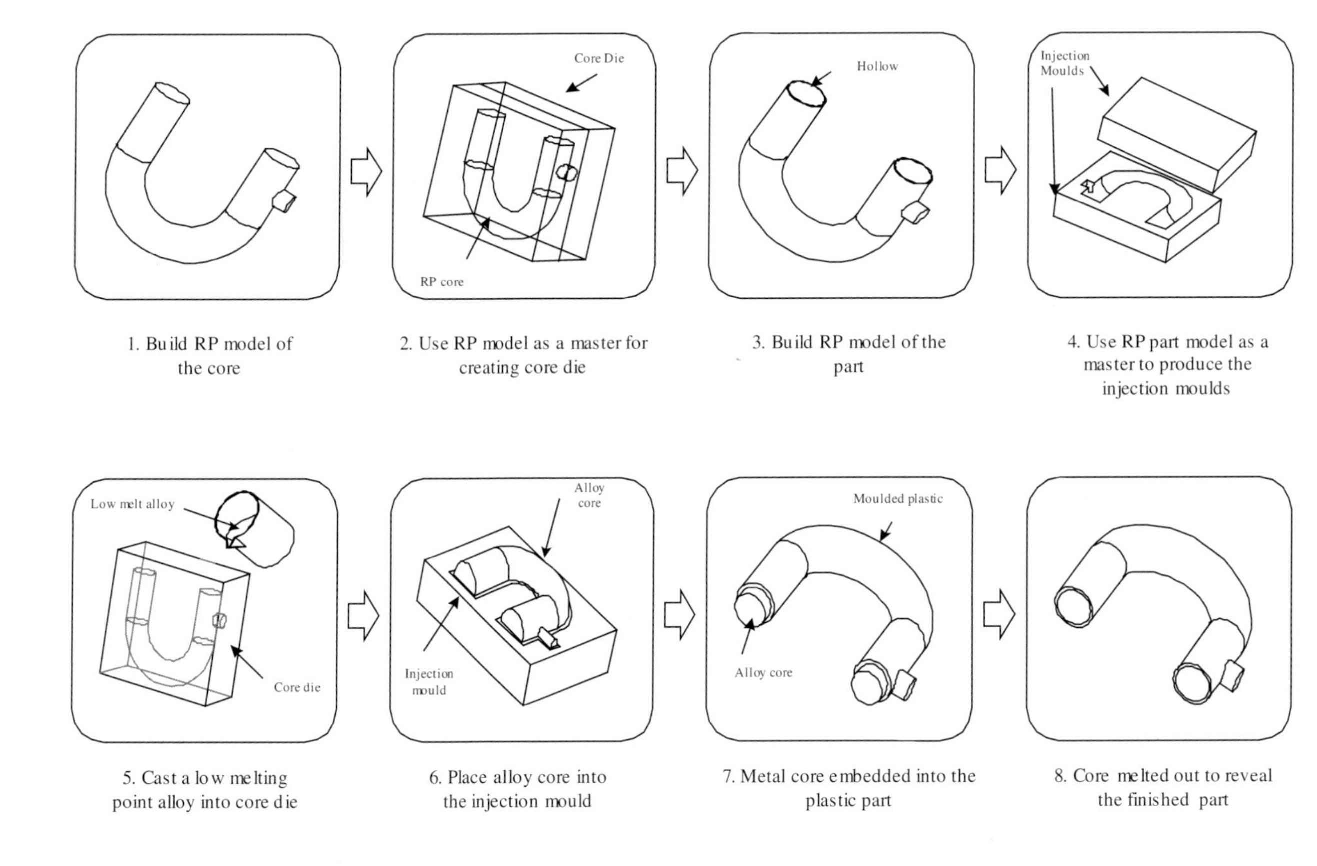

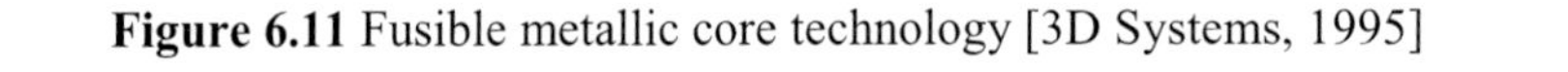

Figure 6.11 Fusible metallic core technology [3D Systems, 1995]

The fusible cores must have the shape of the interior passage of the part. Usually, the core is designed to be placed into a suspended position in the mould and is contained by the mould cavity. The moulded parts encapsulate the removable cores that are melted away by induction heating or by immersing the mouldings in hot water or oil. The processing stages of this technology are shown in Figure 6.11 [3D Systems, 1995]. The cores can have a melting point of up to 220°C depending on the alloys used.

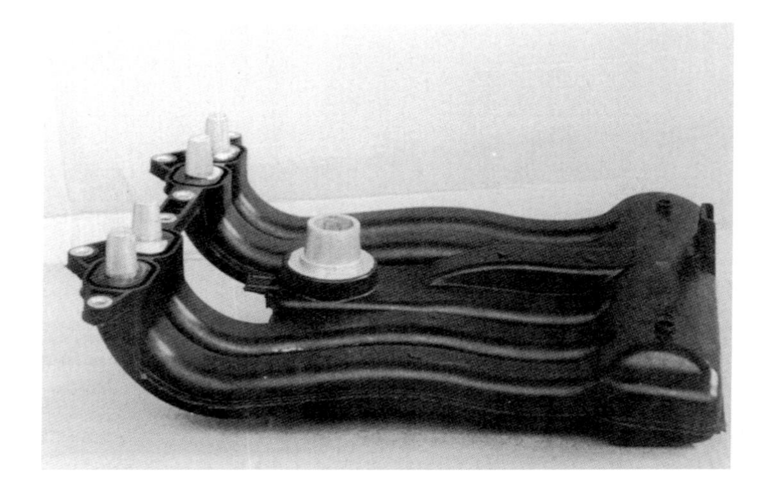

(a) Core encapsulation with thermoplastic

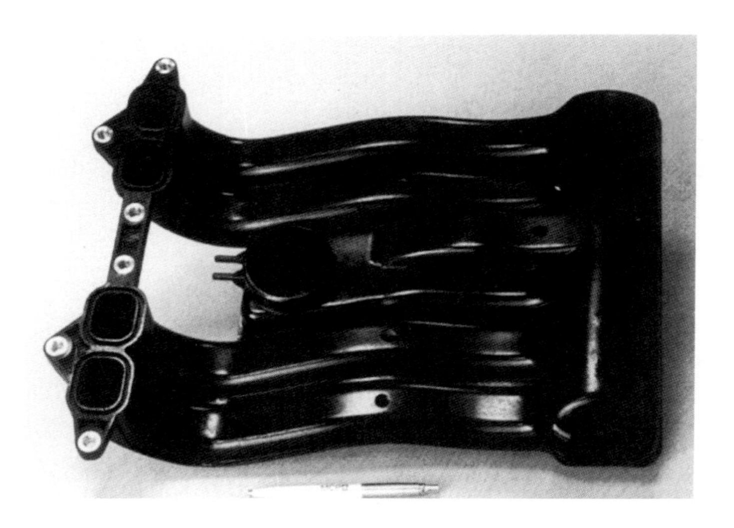

(b) Finished inlet manifold

Figure 6.12 Inlet manifold (Courtesy of MCP)

By using this process, complex parts can be produced without splitting them into pieces and with thin walls that are of a consistent thickness. In this way, it is possible to mould complex parts without welding lines and bolt flanges. Currently, this technology is employed mostly for the production of parts for under-the-bonnet applications. Typical components produced with fusible core technology are: fluid reservoirs, fuel and water pumps, and engine inlet manifolds. An example of a one-piece inlet manifold produced using the technology is shown in Figure 6.12.

6.9 Sand Casting

The sand casting process is often employed for the production of relatively large metal parts with low requirements for surface quality. RP techniques can be utilised to create master patterns for fabricating sand moulds. These moulds are produced by placing RP patterns in a sand box which is then filled and packed with sand to form the mould cavity.

When employing RP techniques, it is much more convenient to build patterns that include compensation for the shrinkage of the castings as well as additional machining stock for the areas requiring machining after casting. The other benefits of employing RP techniques are significantly reduced lead-times and increased pattern accuracy.

6.10 3D Keltool™ Process

The 3D Keltool™ process is based on a metal sintering process which 3M introduced in 1976. This process converts RP master patterns into production tool inserts with very good definition and surface finish.

The production of inserts employing the 3D Keltool™ process involves the following steps [Jacobs, 1996]:

1. Fabricating master patterns of the core and cavity;

2. Producing RTV silicone rubber moulds from the patterns;

3. Filling the silicone rubber moulds with a metal mixture (powdered steel, tungsten carbide and polymer binder with particle sizes of around 5 ηm) to produce "green" parts (powdered metal held together by the polymer binder) duplicating the masters;

4. Firing the "green" parts in a furnace to remove the plastic binder and sintering the metal particles together;

5. Infiltrating the sintered parts (70% dense inserts) with copper in a second furnace cycle to fill the 30% void space;

6. Finishing the core and cavity.

3D Keltool inserts can be built in two materials, Stellite or A-6 composite tool steel. The material properties allow the inserts produced using this process to withstand more than 1,000,000 moulding cycles. Examples of inserts manufactured using the DirectTool™ process are shown in Figure 6.13.

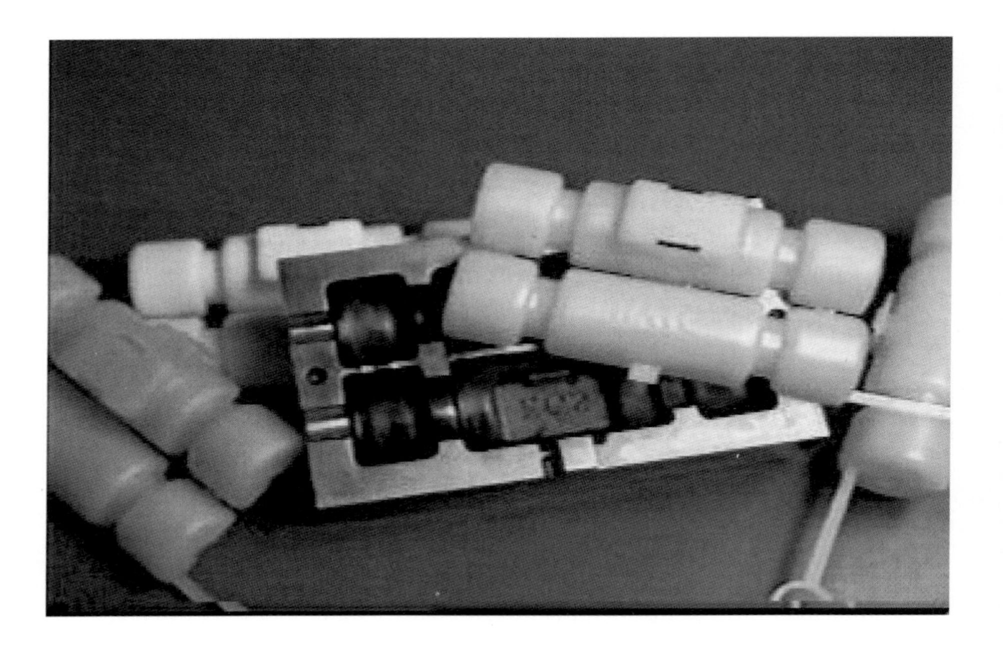

Figure 6.13 3D Keltool inserts (Courtesy of American Precision Products)

6.11 Summary

Indirect tooling methods are intended as prototyping or pre-production tooling processes and not production methods. Consequently, tools fabricated employing these methods will exhibit differences compared to production tools, for example: larger draft angles, simpler part shapes and lower mechanical and thermal specifications. These differences affect the production cycle time, the part mechanical properties and the tool life. However, the aim of these tooling methods is not to replace production tooling but to make only up to a few hundred parts and therefore these tools do not require the strength for a long life. For the same reason, they do not need to be as efficient as production tools and it is justifiable to adopt a longer cycle time per part to compensate for poor thermal conductivity.

References

Bettany S and Cobb RC (1995) A rapid ceramic tooling system for prototype plastic injection, **First National Conference on Rapid Prototyping and Tooling Research**, 6-7 Nov. 1995, Buckinghamshire, UK, ed. G. Bennett, MEP Pub. Ltd., pp 201-210.

Childs THC and Juster NP (1994) Linear Accuracies From Layer Manufacturing, **CIRP Annals**, Vol. 43-1, pp 163-167.

Comeau D and Dobson S (1996) A soft surface tooling method for rapid prototyping, **Society of Plastic Engineers - RETEC**, Pioneer Valley, March 1996: updated November 1996.

Davy D (1996) Century-old chemistry is reborn in moldmaking, **Modern Plastics Magazine.**

Dickens PM, Stangroom R, Greul M, Holmer B, Hon K K B, Hovtun R, Neumann R, Noeken S and Wimpenny D (1995) Conversion of RP models to investment castings, **Rapid Prototyping Journal**, Vol. 1, 4, pp 4-11.

Dickens PM (1996) Rapid Tooling: A review of the alternatives, **Rapid News**, Vol. 4, 5, pp 54-60.

Erickson R (1996) CASTTOOL prototyping for injection molding, where is it going, **Proceedings of the 1996 Wescon Conference**, Anaheim, CA, USA, pp 317-324.

Jacobs PF (1996a) Recent Advances in Rapid Tooling from Stereolithography, White Paper, **3D Systems**, Valencia, California, USA.

Jacobs PF (1996b) Stereolithography and other RP&M technologies, **Society of Manufacturing Engineers - American Society of Mechanical Engineer**.

Kamphuis K and VanHiel B (1996) Rapid tooling for injection molding using cast resin, **Project Progress Report**, Rapid Prototyping and Manufacturing Institute, Georgia Institute of Technology, Atlanta, Georgia.

Maley K (1994) Using Stereolithography to Produce Production Injection Molds, **Annual Technical Conference**, Conference Proceedings, San Fransisco, CA, USA, Vol. 53, 3, pp 3568-3570.

MCP Web Page (2000), **HEK GmbH**, Kaninchenborn 24-28, D-23560 Lübeck, Germany, http://www.mcp-group.de.

Mosemiller L and Schaer L (1997) Combining RP and Spin-Casting, **Prototyping Technology International '97**, UK & International Press, Surrey, UK, pp 242-246.

Mueller T (1995) Stereolithography-Based Prototyping: Case Histories of Application in Product Development, **IEEE Technical Application and Conference Workshops**, Portland Oregon, October 10, pp 305-309.

Plavcan JE (1995) Rapid tooling for compression molding of a thermoset, **Annual Technical Conference**, Conference Proceedings, Boston, MA, USA, Part 3, pp 4324-4326.

Schaer L (1995) Spin-Casting fully functional metal and plastic parts from stereolithography models, **Proceedings of the 6th International Conference on Rapid Prototyping**, Dayton, Ohio, USA, Ch. 27, pp 217-235.

Tritech company literature (1997) **Tritech Precision Products Ltd.**, Wrexham, UK.

Whitward L (1996) Getting to metal quicker with RP patterns, **Design Engineering**, London, January, pp 39-42.

3D Systems Application Guide (1995) **3D Systems**, Valencia, California, USA.

3D Systems White Paper (1996). Recent advances in Rapid Tooling from Stereolithogtaphy, **3D Systems**, Valencia, California, USA.

Chapter 7 Direct Methods for Rapid Tool Production

Indirect methods for tool production as described in the previous chapter necessitate a minimum of one intermediate replication process. This might result in a loss of accuracy and could increase the time for building the tool. To overcome some of the drawbacks of indirect methods, some RP apparatus manufacturers have proposed new rapid tooling methods that allow injection moulding and die-casting inserts to be built directly from 3D CAD models. This chapter describes direct RT solutions that are currently commercially available.

7.1 Classification of Direct Rapid Tooling Methods

Direct RT methods enable the production of inserts capable of surviving from a few dozen to tens of thousands of cycles and represent good alternatives to traditional mould making techniques. The durability or life expectancy of the inserts produced by these methods varies significantly depending on the material and the RT method employed. This makes the application area of direct RT processes also very wide, covering prototype, pre-production and production tooling. According to their application, direct RT processes can be divided into two main groups.

The first group includes less expensive methods with shorter lead times that are appropriate for tool validation before changes become costly. Direct RT methods that satisfy these requirements are called methods for "firm tooling" (also known as "bridge tooling" [Jacob, 1996a]). RT processes for *firm* tooling fill the gap between *soft* and *hard* tooling, producing tools capable of short prototype runs of approximately fifty to a hundred parts using the same material and manufacturing process as for final production parts.

The second group includes RT methods that allow inserts for pre-production and production tools to be built. RP apparatus manufacturers market these methods as "hard tooling" solutions. Currently available solutions for "hard tooling" are based on the fabrication of sintered metal (steel, iron and copper) powder inserts infiltrated with copper or bronze (Keltool™ from 3D Systems, DTM RapidTool™ process, EOSINT Metal from EOS, Three-Dimensional Printing of metal parts from Soligen).

Figure 7.1 shows the classification of direct RT methods according to their application.

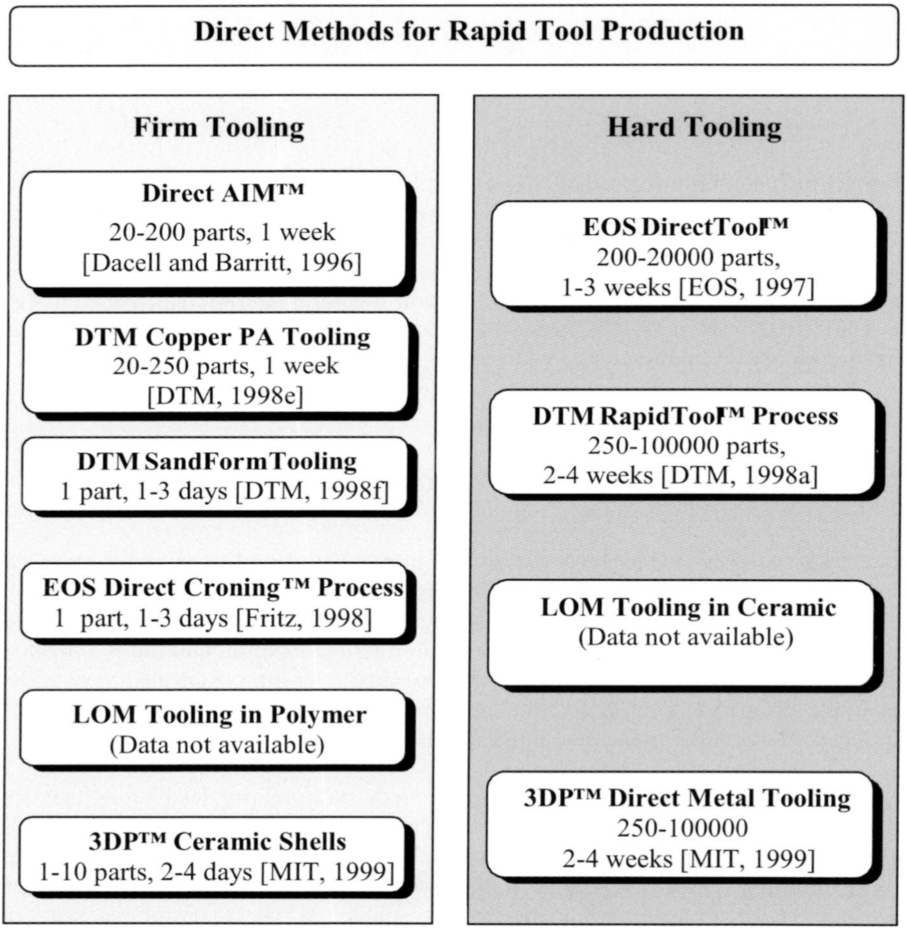

Figure 7.1 Classification of direct RT methods

7.2 Direct ACES™ Injection Moulds (AIM™)

Stereolithography is used to produce epoxy inserts for injection mould tools for thermoplastic parts. Because the temperature resistance of the curable epoxy resins available at present is up to 200°C (Cibatool®SL5530HT) and thermoplastics are injected at temperatures as high as 300°C (572°F) specific rules apply to the production of this type of injection mould. The procedure detailed in [Decelles and Barritt, 1996] is outlined below (Figure 7.2).

Using a 3D CAD package, the injection mould is drawn. Runners, fan gates and ejector pin clearance holes are added and the mould is shelled to a recommended thickness of 1.27mm (0.05"). The mould is then built using the Accurate Clear Epoxy Solid (ACES) style [Jacobs, 1996b] on a stereolithography machine. The supports are subsequently removed and the mould is polished in the direction of the draw to facilitate part release. The thermal conductivity of the stereolithography resins is about 300 times lower than that of conventional tool steels (0.2002 W/mK for Cibatool®SL5170 epoxy resin) [Jacobs, 1996a]. To remove the maximum amount of heat from the tool and reduce the injection moulding cycle time, copper water cooling lines are added and the back of the mould is filled with a mixture made up 30% by volume of aluminium granulate and 70% of epoxy resin. The cooling of the mould is completed by blowing air on the mould faces as they separate after the injection moulding operation.

The main disadvantage of Direct AIM™ (ACES™ Injection Moulds) is that the number of parts that can be obtained using this process is very dependent on the shape and size of the moulded part as well as the skills of a good operator who can sense when to stop between cycles to allow more cooling. Because finishing must be done on the internal shapes of the mould, the process is slightly more difficult than for indirect methods where most of the model shapes are external. Also, draft angles of the order of ½ to 1 degree and the application of a release agent in each injection cycle are required to ensure proper part ejection.

A Direct AIM™ mould is unlikely to be as durable as an aluminium filled epoxy mould. The backing operation influences the cooling time but has no beneficial effect on the life of the mould which is subject to erosion by the injected material.

Although the injection cycle time (3 to 5 min) is long by comparison with conventional injection moulding (5 to 15 s), a Direct AIM™ tool suitable for moulding up to 100 parts can be completed and the parts produced in the preferred material within a week of receiving the design.

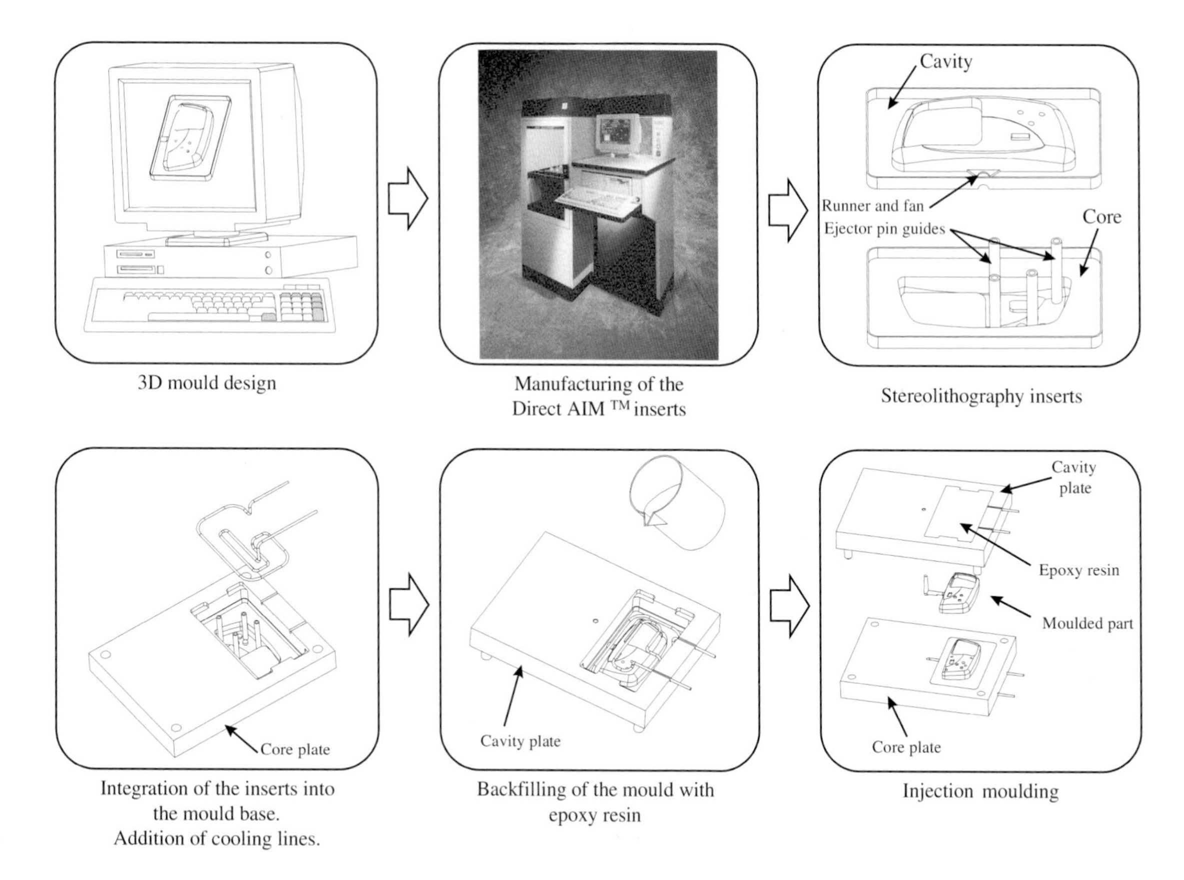

Figure 7.2 Direct AIM™ injection mould

To increase both the resistance to erosion and the thermal conductivity of Direct AIM™ tools, the deposition of a 25μm layer of copper on the mould surface has been investigated [Jacobs, 1996a].

Similar injection moulds made of nylon and produced by SLS have also been reported [Venus and Van de Crommert, 1996], as well as injection moulds created directly by stereolithography but using urethane photocurable resin containing a high level of glass particles [Nakagawa, 1994].

7.3 Laminated Object Manufactured (LOM) Tools

The original LOM process produces parts with a wood like appearance using sheets of paper. Experiments to build moulds directly or coated with a thin layer of metal have been reported [Pak et al., 1997]. Unfortunately, moulds built this way can only be used for low melting thermoplastics and are not suitable for injection moulding or blow moulding of common thermoplastics. For this reason, new materials based on epoxy or ceramic capable of withstanding harsh operating conditions have been developed.

- Polymer sheets: These sheets consist of glass and ceramic fibres in a B-staged epoxy matrix. Parts made with this material require postcuring at 175°C for one hour. Once fully cured, they have good compressive properties and a heat deflection temperature of 290°C [Klosterman et al., 1996].

- Ceramic sheets: Two ceramic materials have been developed for LOM, a sinterable AlN ceramic and a silicon infiltratable SiC ceramic. Both materials are mixed with 55% by volume of polymeric binder [Klosterman et al., 1996].

The polymer composite process is being beta-tested and the first industrial application is expected soon. The ceramic process is less advanced and requires more software and hardware modifications to the LOM machine. Few results for these processes are available but current indications are promising.

7.4 DTM RapidTool™ Process

SLS is one of the rapid prototyping techniques widely used for direct tool production. Using SLS, DTM was one of the first companies to commercialise a rapid tooling technology, marketing it as the RapidTool™ process. The DTM RapidTool family of tooling products consists of three materials, RapidSteel 1.0, RapidSteel 2.0 and Copper Polyamide. Each of these materials requires different processing techniques although all three materials are designed for use with the same DTM Sinterstation platform (see Chapter 3). In addition to the RapidTool family of materials, DTM has developed a foundry sand (SandForm^{TM}) for direct fabrication of moulds and cores for sand casting tooling.

7.4.1 RapidSteel 1.0

The first product, RapidSteel 1.0 powder, is made up of low-carbon steel particles with a mean diameter of 55μm. The particles are coated with a thermoplastic binder. The processing of RapidSteel 1.0 described in Figure 7.3 can be broken down into three main stages [Pham et al., 2000]:

1. *Green part manufacture (SLS processing)*: The low melting point binder allows the material to be processed in the SLS machine without heating the feed and part bed. Tooling inserts in the "green" stage are built layer by layer through fusion of the binder.

2. *Cross-linking*: During the subsequent furnace cycle, the thermoset binder coating would melt and would behave as a lubricant between the steel particles. To prevent distortion being caused in this way during the low temperature portion of the furnace cycle, the green part is infiltrated with an aqueous acrylic emulsion and dried in an oven at about 60°C. The acrylic emulsion acts as a binding agent that provides strength to the green part when the polymer is burnt away in the furnace. The drying time is dependent upon the part size; for large parts, it can take up to 48h.

3. *Furnace processing*: In this stage, the green part is converted into a fully dense metal part by infiltration with molten copper. To remove oxides from the steel surface, a mixture of hydrogen and nitrogen is used during the furnace cycle. Between 350 and 450°C, the polymer evaporates. Then, the temperature is increased to 1000°C to allow the sintering of the steel powder. Finally, the part is heated up to 1120°C where copper infiltration occurs driven by capillary action.

The final RapidSteel 1.0 parts (Figure 7.4) are 60% steel 40% copper fully dense parts which can be finished by any technique including surface grinding, milling, drilling, wire erosion, EDM, polishing and surface plating.

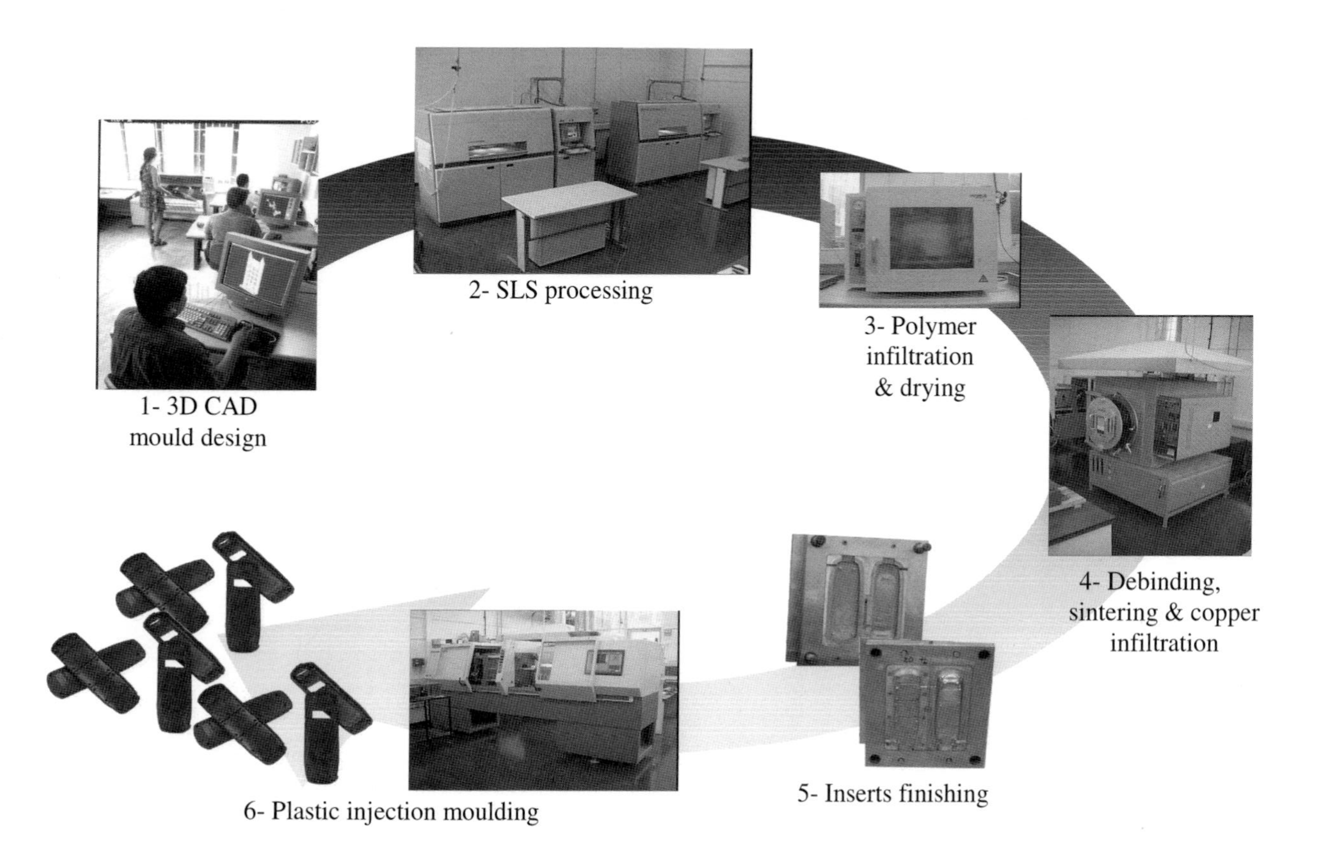

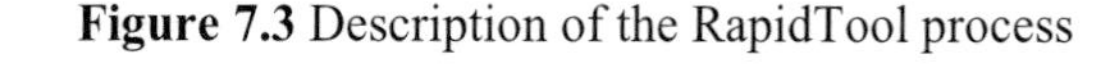

Figure 7.3 Description of the RapidTool process

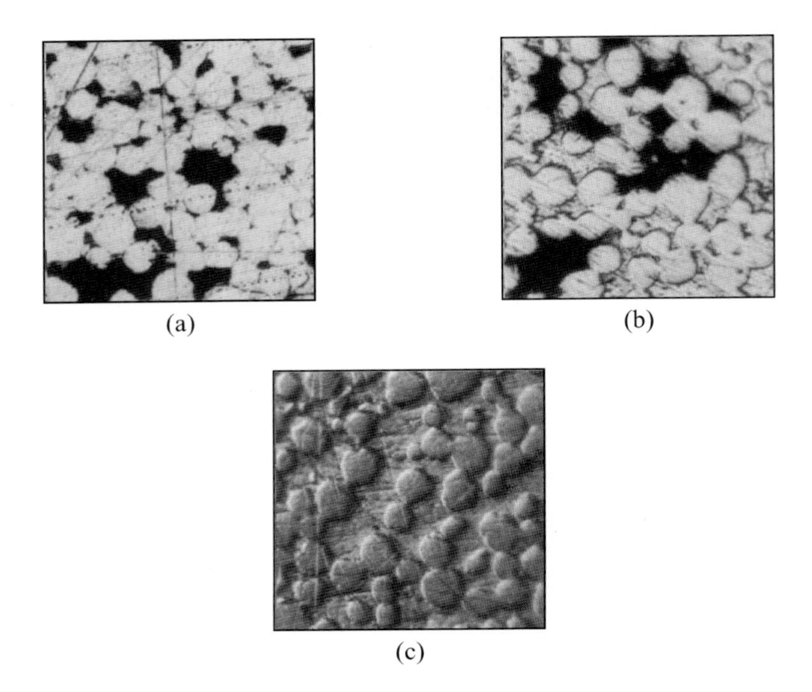

Figure 7.4* Optical micrograph of a RapidSteel part surface
(a - Sintered Part, b - Partially copper infiltrated, c- Fully copper infiltrated)

There is no restriction on the complexity of the geometry of the part that can be produced by SLS. However, the RapidTool process imposes a number of constraints on insert design:

- Excess material has to be added on the parting surfaces, shut-offs and sides of the inserts and machined afterwards for a good surface match.
- The process requires a flat base for proper infiltration.
- In some cases, it is useful to omit some simple features from the insert design if they can be easily machined or inserted later. This is particularly recommended for features smaller than 1mm that are difficult to build and can easily be broken or damaged during the cleaning of the green part. Additionally, the offset value used for the laser beam can be as high as 0.4 mm which also limits the maximum resolution achievable.
- Features that require a very high positioning accuracy or tight tolerances and that are easy to machine, eg holes, or that can be easily added, eg bosses, have to be removed from the CAD model before the SLS process.

- In the furnace, between the debinding and sintering stages, the part shape is maintained only by friction between the steel particles. Features such as holes across the whole part or cooling lines have to be avoided because they can weaken it, causing it to distort and even to collapse. During each of the RapidTool process steps, parts are subject to size variations. Figure 7.5 shows the distribution of dimensional changes occurring during the process.

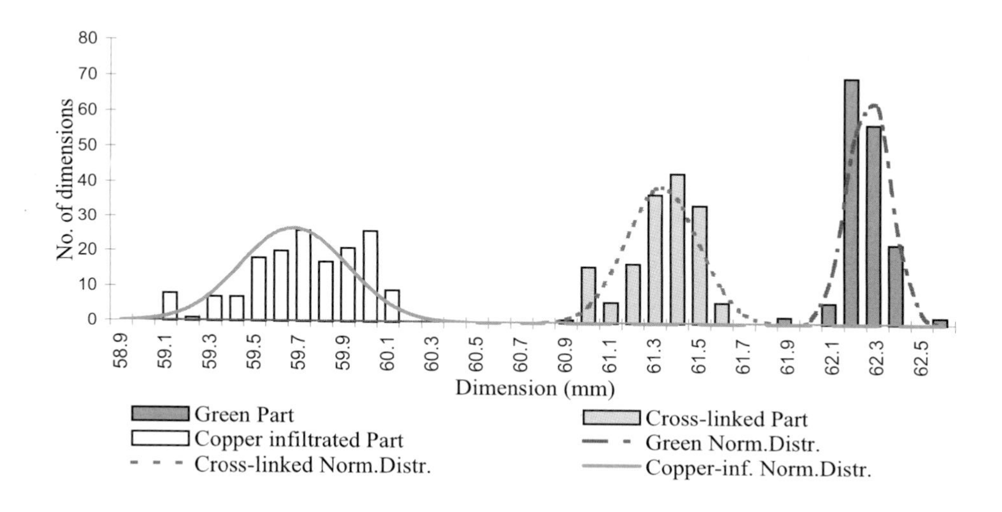

Figure 7.5 Dimensional variations for a nominal dimension of 60 mm

The dimensional accuracy decreases after each step of the process as the variance of the distribution increases. These variations affect not only part dimensions but sometimes the part geometry can also change. Although more shrinkage occurs during the furnace cycle, for all stages, the drying of the acrylic is most problematical because the surface of the part dries more quickly than the inside, which prevents the inside from shrinking freely. This introduces internal stresses and part distortion. Shrinkage is assumed to be linearly related to part dimension and is compensated for by the application of scaling factors to the part prior to the build. However, experience shows that the shrinkage can vary from one part to another as it is influenced by factors such as the weight and geometry of the part. By first building a part to calculate scaling factors and offset values, an accuracy of ±0.1% has been achieved [Reinhart and Breitinger, 1997]. However, this method increases the time and cost of processing parts. Also it cannot be repeated for every part. Unfortunately, even with experience, part shape variations are not always fully predictable and the best way to increase part accuracy would be to reduce significantly the shrinkage occurring during the whole process. In a previous paper [Pham et al., 1999], the authors have described how the geometry, size and weight of the part influence part accuracy. They have proposed to compensate for this by tuning the scaling factors and using different offset values depending on whether the part has a convex or concave geometry. The accuracy of the process as claimed by the manufacturer is ±0.25%, but experience shows that realistically IT 16 is more achievable.

Correct finishing of the tool is also important to obtain better accuracy for the moulded parts. Usually, the best surface to adopt as a reference for the machining of the inserts is the bottom plane as it is flatter and less rough than the vertical sidewalls (roughness value of 8.2 µm Ra as compared to 10.8 µm Ra). The roughness of freeform surfaces is very dependent on the slope. However, the roughness expected from the stair stepping effect is usually decreased by the copper which creates a meniscus between the steps. The slight erosion by brushes or compressed air during the cleaning of the part also tends to reduce roughness. The finishing and polishing of RapidTool inserts is mainly a manual process that can be laborious and time consuming. However, it can be significantly speeded up by initial finishing of the inserts before the furnace cycle. After polishing, a surface finish of 0.3 µm Ra can be obtained. Examples of inserts produced using RapidSteel 1.0 are shown in Figure 7.6.

7.4.2 RapidSteel 2.0

RapidSteel 2.0 (announced in May 1998) [DTM, 1998a] offers a number of modifications over RapidSteel 1.0:

- The base metal has been changed from carbon steel to 316 stainless steel.
- Bronze has replaced copper as the infiltrant.
- The thermoplastic binder material has been substituted by a thermoset binder.

As a result of these changes, the processing of RapidSteel 2.0 differs from that of RapidSteel 1.0, and comprises the following stages:

1. Green part manufacture (SLS processing): The part bed is heated to a temperature of about 100°C for the SLS processing.
2. Brown part manufacture (Furnace debinding and sintering cycle): The temperature is raised to 1120°C, held for 3 hours and decreased to room temperature in a controlled atmosphere.
3. Part infiltration (Furnace bronze infiltration cycle): The temperature is raised to 1050°C, held for 2 hours and decreased to room temperature in a controlled atmosphere.

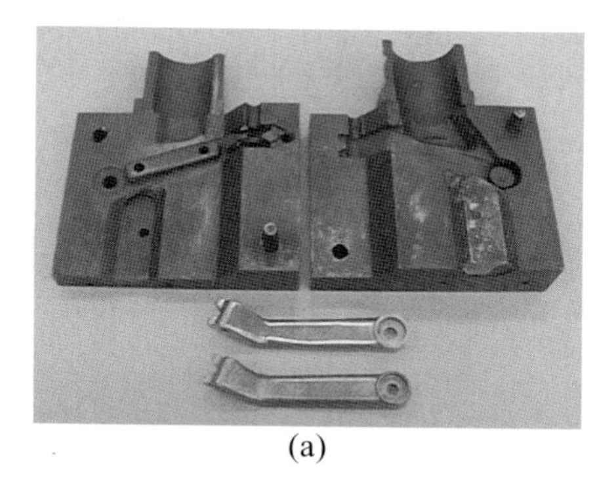

(a)

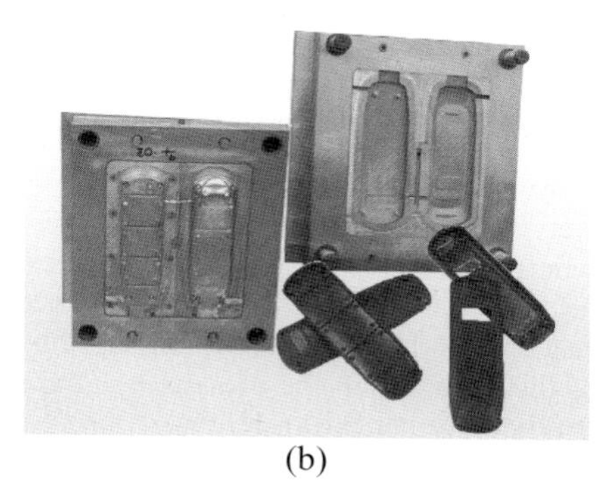

(b)

Figure 7.6 Inserts built using RapidSteel 1.0
(a - windscreen wiper arm die-casting inserts and parts, b – electronic tour guide front and back covers inserts.

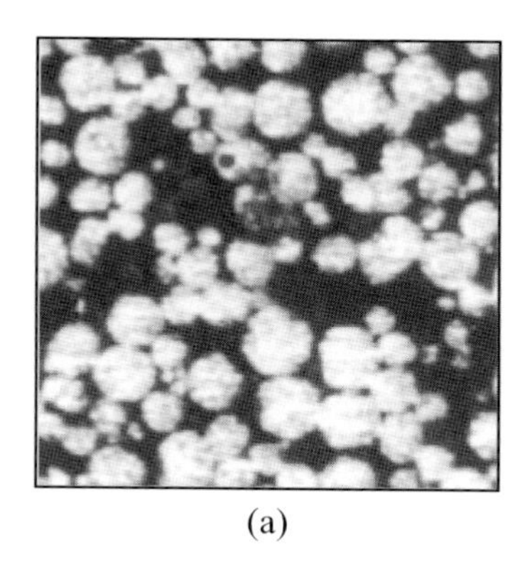

(a)

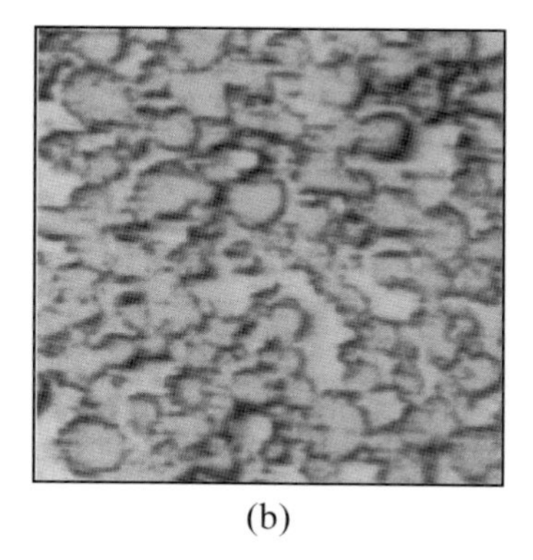

(b)

Figure 7.7* Optical micrograph of a RapidSteel 2.0 part surface
(a - Sintered part, b - Fully bronze infiltrated part)

The final RapidSteel 2.0 parts (Figure 7.7) are made up of 53% stainless steel 47% bronze [DTM, 1998b] and can also be finished by any conventional technique.

A comparison of the main properties of RapidSteel 1.0 and RapidSteel 2.0 is given in Figure 7.8.

From initial experience, it is possible to state that the decrease of the average particle size from 55µm to 34µm allows parts to be built with a smaller layer thickness of 75µm. This leads to smoother surfaces through a reduction of stair stepping effects and consequently shortens the time required for finishing. The particle size reduction also increases the part resolution by allowing the formation of sharper edges. However, the minimum feature size still remains in the order of 1mm. The reason for this is that, although the laser beam diameter is small, the thermal conductivity of the steel powder causes the beam to sinter an area wider than its diameter.

The properties of bronze make the part easier to polish and give better friction characteristics to the mould inserts. In addition, the change of the metal base from carbon steel to 316 stainless steel should increase the mould wear resistance.

Although the process now requires two furnace cycles due to the sintering temperature being higher than the infiltration temperature, each of the cycles is about 20h which is less in total than the initial 48h furnace cycle for RapidSteel 1.0. The modification of the binder material eliminates the need for the cross-linking stage. This reduces considerably the total processing time by saving up to 48h for large parts and increases the accuracy of the process by eliminating the distortion of parts during the drying of the polymer infiltrant.

A new test part has been introduced and internal and external features are taken into account to calculate the shrinkage factors and the offset values. However, the accuracy of RapidSteel 2.0 seems less sensitive to part geometry and weight mainly because of the elimination of the cross-linking stage and the reduction in dimensional changes. Instead of a shrinkage of about 4-5%, the estimated change for the whole process is an expansion of 0-0.5%. This explains the lower percentage of steel after infiltration although the size of the particles has decreased.

Figure 7.9 shows the achievable average dimensional accuracy of parts produced in RapidSteel 1.0 and RapidSteel 2.0 materials. Initial investigations have shown that the accuracy has greatly improved and is now much closer to the ±0.1mm usually required for production injection moulding tools. Figure 7.10 shows examples of inserts manufactured using RapidSteel 2.0.

	Units (SI)	Test Method	RapidSteel 1.0	RapidSteel 2.0
Physical Properties				
Average Particle Size	μm		50	34
Density	g/cm^3	ASTM D792	8.23 (@20°C)	7.5 (@23°C)
Thermal Properties				
Thermal Conductivity	W/mK	ASTM E457	185 (@100°C)	23 (@100°C) 28 (@200°C)
Coefficient of Thermal Expansion	μm/mK	ASTM E831	14.4 (51 to 232°C)	14.6 (51 to 150°C)
Mechanical Properties				
Yield Strength (0.2%)	MPa	ASTM E8	255	413
Tensile Strength	MPa	ASTM E8	475	580
Tensile Elongation Break	%	ASTM E8	15	0.9
Young's Modulus	GPa	ASTM E8	210	263
Hardness		ASTM E18	75 Rockwell "B"	22 Rockwell "C"

Figure 7.8 RapidSteel 1.0 and RapidSteel 2.0 properties [DTM, 1998c]

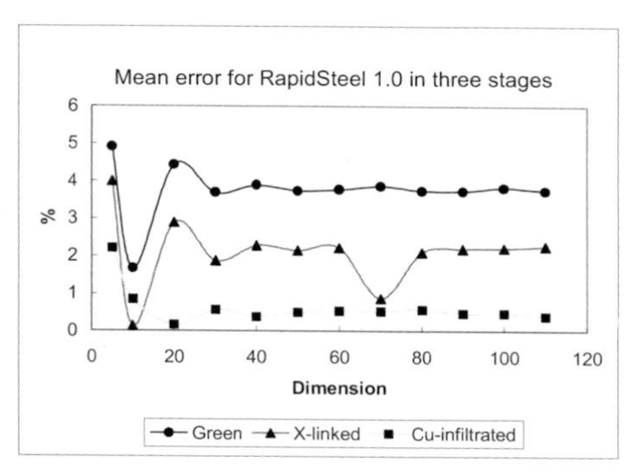

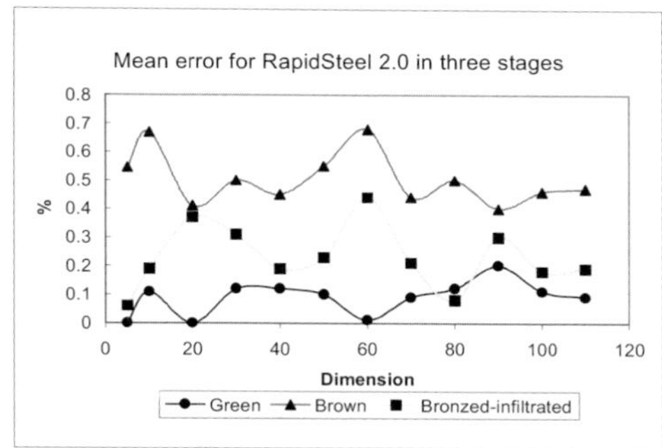

Figure 7.9 Average errors from all processes

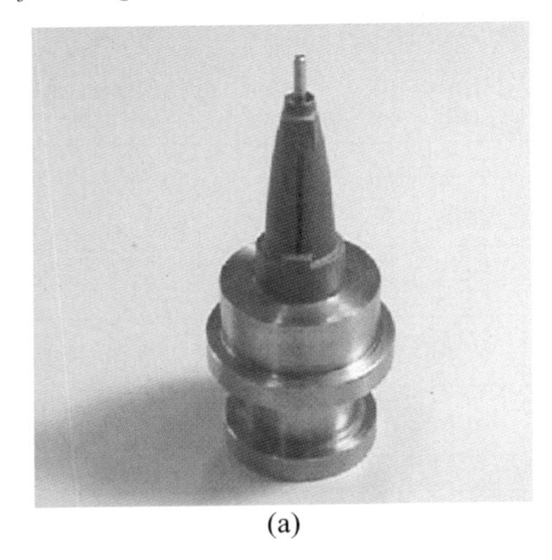

(a)

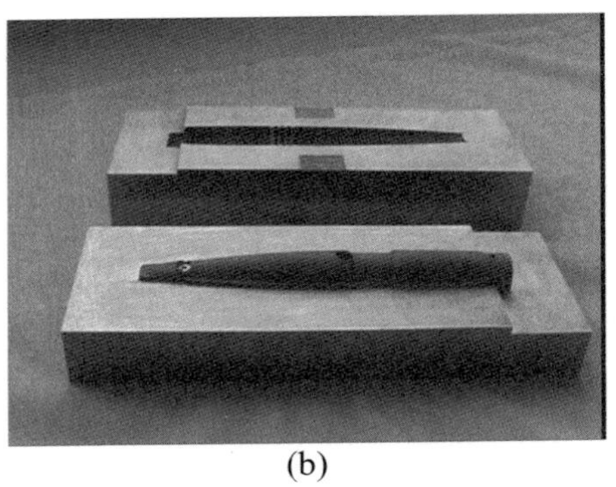

(b)

Figure 7.10 Set of inserts for a nose hair trimmer

7.4.3 Copper Polyamide (PA)

Copper PA (announced in May 1998) [DTM, 1998d, DTM, 1998e] is a new metal-plastic composite designed for short-run tooling applications involving several hundred parts (100-400 parts) from common plastics. Tooling inserts are produced directly in the SLS machine with a layer thickness of 75μm and only subsequent finishing is necessary before their integration in the tool base. No furnace cycle is required and unfinished tool inserts can be produced in a day.

During the CAD stage, Copper PA inserts are shelled and cooling lines, ejector pin guides, gates, and runners are included in the design and built directly during the SLS process (Figures 7.11 and 7.12). Then, the insert surfaces are sealed with epoxy and finished with sandpaper, and finally the shell inserts are backed up with a metal alloy. Examples of parts moulded using Copper PA inserts are shown in Figure 7.13.

Inserts produced from Copper PA are easy to machine and finish. Their heat resistance and thermal conductivity are better than most plastic tooling materials [DTM, 1998d and 1998e]. The cycle times of moulds employing Copper PA inserts are similar to those for metal tooling.

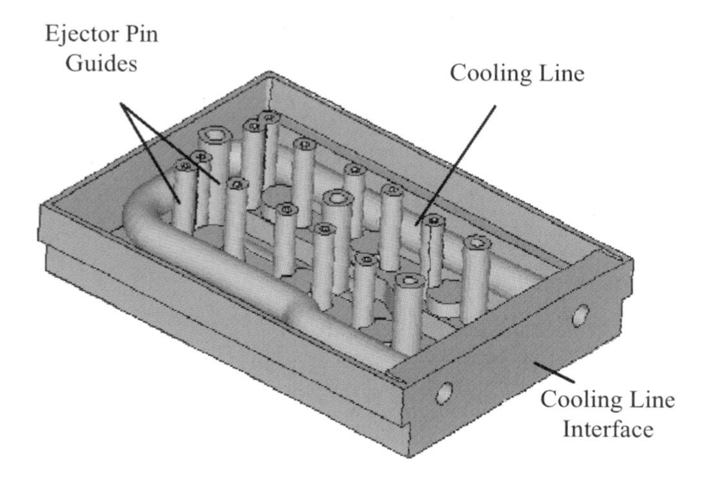

Figure 7.11 Back surface of an insert [DTM, 1998e]

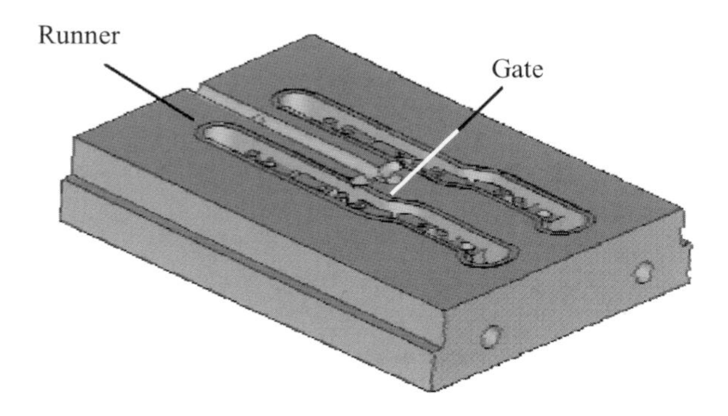

Figure 7.12 Working surface of an insert [DTM, 1998e]

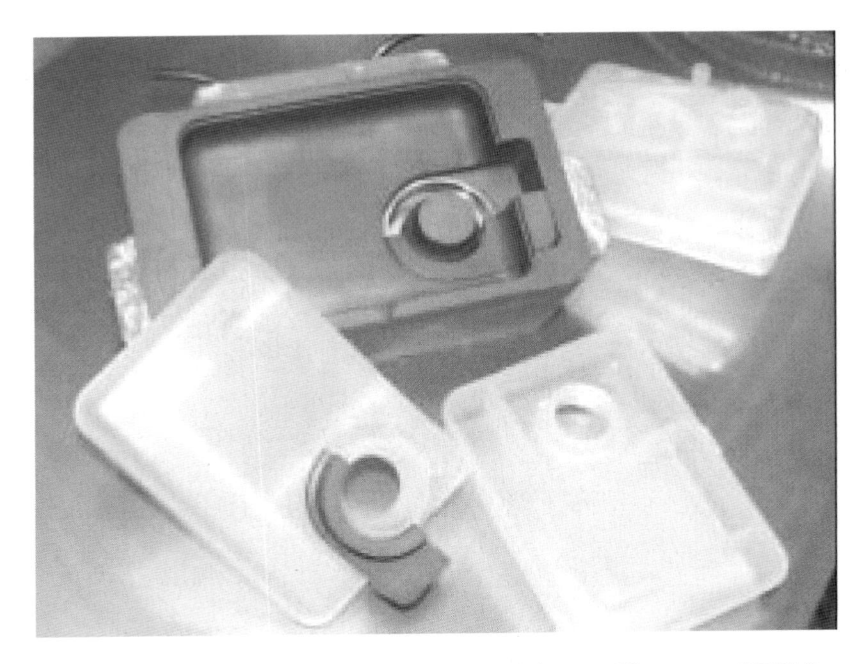

Figure 7.13 Parts moulded using Copper PA inserts (Courtesy of DTM)

7.5 SandForm™

SandForm™ Zr & Si materials can be used to build moulds and cores directly from 3D CAD data employing the SLS process. The sand moulds and cavities produced are of equivalent accuracy, and have properties that are identical to cores fabricated with conventional methods. SandForm™ moulds and cores can be used for low-pressure sand casting. Figure 7.14 shows the core of a aircraft engine part produced by DTM for an industrial client [DTM, 1998f].

7.6 EOS DirectToolTM Process

DirectTool™ is a commercial rapid tooling process introduced by EOS GmbH in 1995. Ongoing material and process developments have increased the process productivity and the quality of the built parts. The most common application of this process is the production of inserts for plastic injection moulding and rubber vulcanisation. This direct tooling process uses special proprietary metal powders which are selectively sintered in a specially developed machine, the EOSINT M

250. The sintered parts are porous and usually undergo infiltration with an epoxy resin to increase their strength [Fritz, 1998]. After infiltration, further polishing of the part surfaces is possible to achieve the quality required for injection moulding inserts.

Figure 7.14* SandForm™ core (Courtesy of DTM)

In 1999, EOS introduced their second generation DirectTool™ process together with the new EOSINT M 250 X^{tended} machine (Figure 7.15). In addition to materials with low melting points, the new machine can process steel powder for the first time.

The DirectTool™ process is mainly utilised for rapid tooling of complex inserts the surfaces of which cannot be machined directly. The process is considered a viable alternative for prototype and pre-production tooling applications requiring the manufacture of up to a few thousands parts from common engineering plastics.

Building envelope	250 x 250 x 185 mm3
Laser	CO^2, min. 200 W
Laser scan speed	up to 3 m/s
Building speed	2-15 mm^3/s (material dependent)
Layer thickness	0.05-0.1 mm
Process computer	PC
Electrical supply	400 V, 32 A
Dimensions	1950 x 1100 x 1850 mm^3
Weight	900 kg
Workstation	Silicon Graphics Indigo
PC	Windows 95, Windows NT
Interface to CAD	Standard: STL, CLI Optional: VDA-FS, IGES, CATIA

Figure 7.15 EOSINT M 250 X^{tended} Specifications [EOS, 1997]

The new powders, DirectSteelTM 50-V1 and DirectMetalTM 50-V2, allows inserts with layer thickness of 50 μm to be built enabling reproduction of intricate structures and details [Fritz, 1998]. EOS has introduced another powder, DirectMetalTM 100-V3, with a maximum particle size of 100 μm, for use at higher building speeds. Examples of inserts produced using the DirectToolTM process are shown in Figure 7.16. The inserts were built from the DirectSteel™ 50-V1 material and a series of plastic parts were successfully injection moulded in a highly abrasive 50% glass-fibre reinforced polyamide [EOS, 1997].

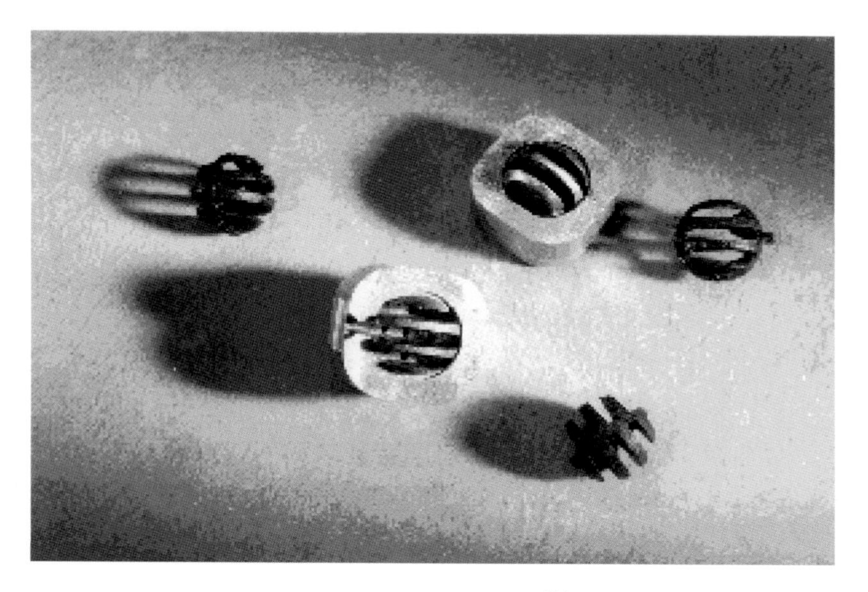

Figure 7.16 Inserts produced using the DirectTool™ process (Courtesy of EOS)

In addition to the DirectTool™ process, EOS has developed a system, EOSINT S, for the production of sand casting moulds and cores by directly sintering resin-coated foundry sand [Fritz, 1998]. This technology is also known as the Direct Croning Process (DCP). The moulds and cores produced by this process can be used to cast metal prototypes or small series of production parts in all conventional sand-casting materials. The process is compatible with the standard production Croning process. Figure 7.17 shows a car cylinder head produced using the EOSINT S system.

Figure 7.17 Car cylinder head produced using EOSINT S System [Fritz, 1998]

7.7 Direct Metal Tooling using 3DP™

The 3D Printing process developed by the MIT [Sachs et al., 1997; MIT, 1999] can be employed to build metal parts for injection moulding tooling inserts from a CAD model in a range of materials including stainless steel, tungsten and tungsten carbide. The process allows the fabrication of parts with overhangs, undercuts and internal volumes as long as there is an escape route for the unused loose powder. The production of metal parts includes the following steps:

1. Building the part by combining powder and binder employing the 3DPTM process.
2. Sintering the printed parts in a furnace to increase their strength.
3. Infiltration of the sintered parts with low melting point alloys to produce fully dense parts.

The 3DP™ process can be easily adapted for production of parts in a variety of material systems, for example metallic/ceramic compositions with novel material properties [Sachs et al., 1997; MIT, 1999]. Tooling inserts built using this process are shown in Figure 7.18.

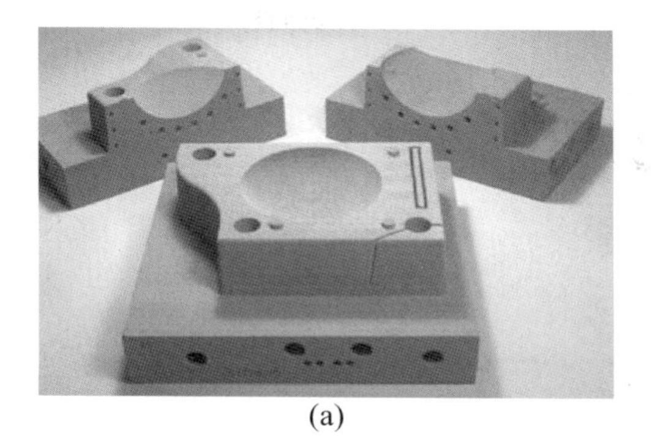

(a)

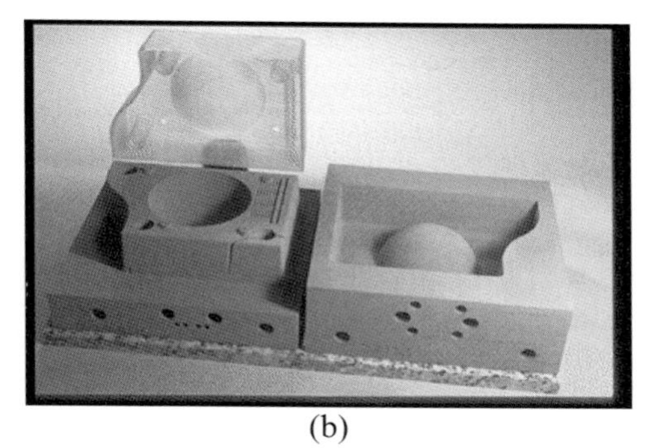

(b)

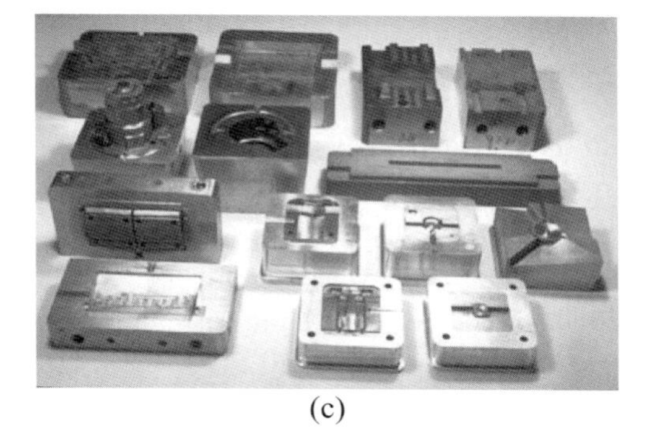

(c)

Figure 7.18 Tooling inserts fabricated using the 3DP™ process (Courtesy of MIT) (a and b - injection moulding inserts with conformal cooling, c - finished metal inserts)

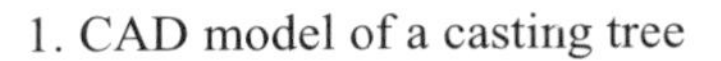

1. CAD model of a casting tree 2. CAD model of a ceramic mould

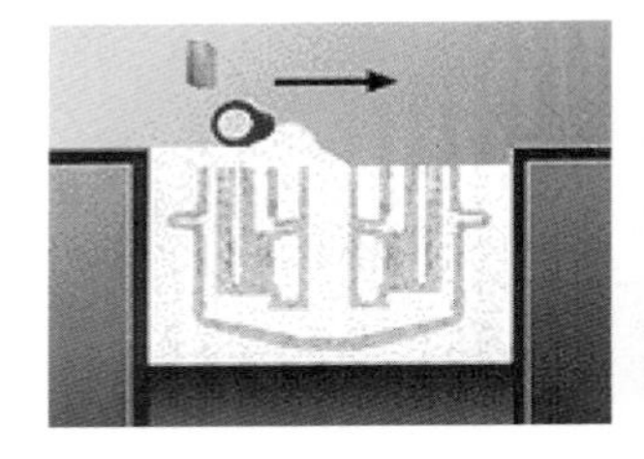

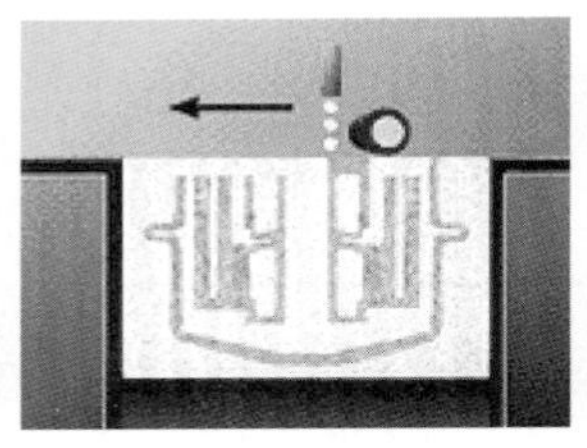

3. Building the mould using 3D printing process

4. Removal of the unbound powder 5. Casting the mould

Figure 7.19* Direct Shell Production Casting (Courtesy of Soligen)

The 3DP™ process is also employed to build ceramic moulds directly from 3D CAD data without any intermediate steps. This process is marketed by Soligen Inc. and is known as Direct Shell Production Casting (DSPC). The DSPC process involves the following steps (Figure 7.19) [Uziel, 1997; Soligen, 1999]:

1. A 3D CAD model is created of a casting tree that includes the gating system through which molten material will flow.
2. The CAD model of the tree is used as a reference to generate a digital model of a ceramic mould.

3. The CAD model of the mould is used to build the actual ceramic mould with the 3DP™ process.

4. The unbound powder is removed from the mould.

5. The mould is filled with molten metal. After solidification of the metal, the ceramic and gating metal are removed and the casting is finished.

Examples of ceramic moulds and castings fabricated using this process are shown in Figure 7.20. The DSPC process has been employed successfully for the production of metal castings for a wide range of applications including aerospace, medical implant and automotive.

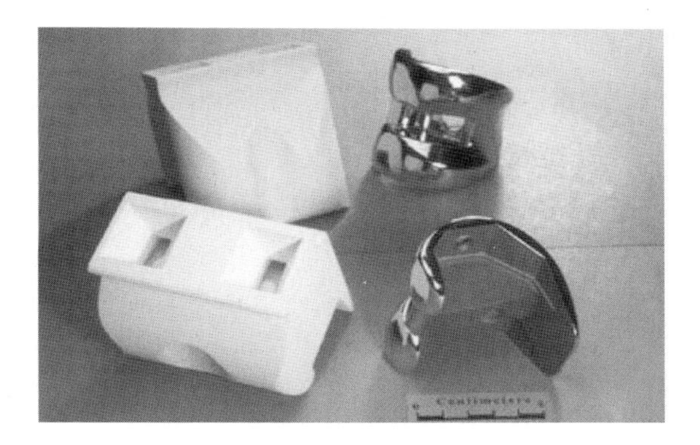

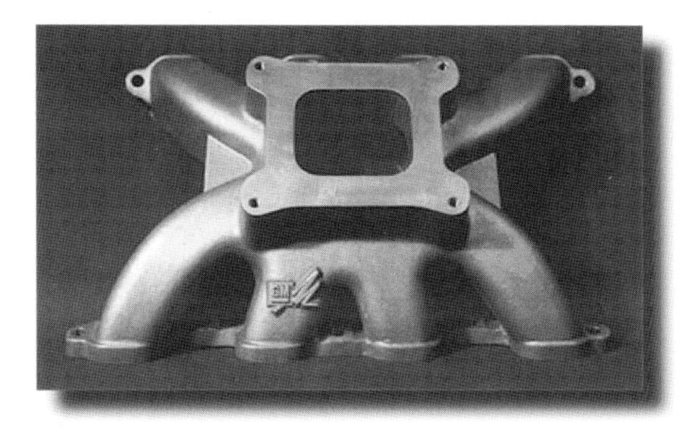

Figure 7.20* Examples of ceramic moulds and castings fabricated using DSPC (Courtesy of Soligen)

7.8 Topographic Shape Formation (TSF)

Topographic Shape Formation (TSF) is very similar to 3DP. This technology is used primarily for rapid production of moulds. The parts are built by successive layering of a silica powder and selective spraying of paraffin wax from an X-Y-Z controlled nozzle (Figure 7.21). The wax binds the powder to form each cross-section of the part and also partially melts the previous layer to ensure good adhesion. Once the part is completed, it is sanded, coated in wax and then employed as a mould for the customer's part. Materials in use include concrete, fibreglass and expanding foam [Formus, 2000].

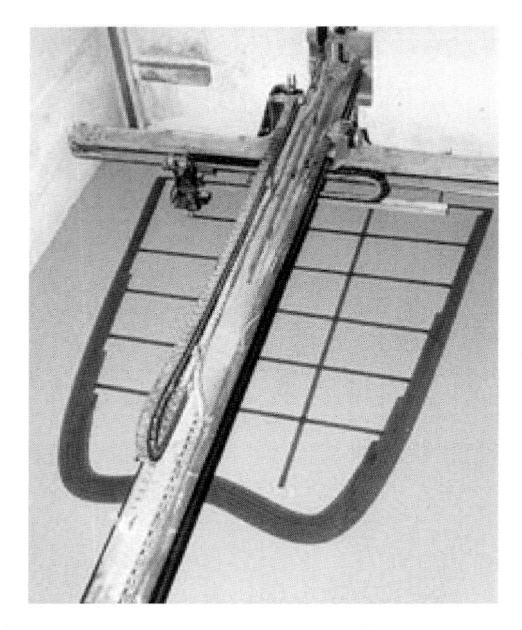

Figure 7.21 Topographical Shape Formation system (Courtesy of Formus)

An advantage of this technology is that it can build very large parts quickly and cheaply, that may be expensive and time-consuming if constructed by other RP methods. A disadvantage is that the moulds have a 'gritty' surface finish and may need to be finished by an operator.

At present, TSF is in use in only one service bureau. The machine has a maximum build envelope of 3353 x 1829 x 1219 mm, a layer thickness of 1.27-3.81 mm, a resolution of 12.7 mm and an accuracy of ±1.27 mm [Formus, 2000].

7.9 Summary

Direct methods for tool production reduce the total production time and the inaccuracies introduced by intermediate replication stages. The restricted range of materials available is still the most severe drawback of direct tooling methods, but materials are improving continuously and new materials such as ceramics for stereolithography are becoming available [Dunlop, 1995]. Another promising direction for further improvement of direct tooling methods is to combine their capabilities with those of traditional tooling methods. In this way, the application area of direct tooling methods can be extended significantly.

References

Decelles P and Barritt M (1996) Direct AIM™ prototype tooling, Procedural Guide, **3D Systems**, Valencia, CA, USA.

Dunlop RN (1995) Physical characteristics of metal sprayed tooling, **First National Conference on Rapid Prototyping and Tooling Research**, 6-7 November 1995, Buckinghamshire, UK, ed. G. Bennett, MEP Pub. Ltd., pp 249-256.

DTM Press Release (1998a) RapidSteel 2.0® Reduces Typical Build Times by Half, Improves Surface Finish in Applications for Metal Mold Inserts, **DTM Corporation,** May 1998, http://www.dtm-corp.com/News/newsdesk.htm#Press.

DTM (1998b) **DTM 2nd European User Group Meeting** Leuven, Belgium, RapidSteel 2.0, 7-8 October 1998.

DTM Product information - RapidTool 2.0 Material (1998c) **DTM GmbH**, Otto-Hahn-Str. 6, D-40721 Hilden.

DTM Press Release (1998d) DTM Targets Short-Run Tooling Applications with new Copper Polyamide Material, May 1998, **DTM Corporation,** http://www.dtm-corp.com/News/newsdesk.htm#Press.

DTM Product information (1998e) Copper Polyamide Mold Insets for Plastic Injection Molding, **DTM Corporation**, June, 81611 Headway Circle, Building 2, Austin, TX.

DTM SandForm™ material fuels development of new aircraft engine part for Woodward Governor company (1998f) **Horizons**, newsletter published by DTM.

EOS Press Release (1997) DMLS and EOSINT M 250 on Stage at EuroMold 1997 December, **EOS**, www.eos-gmbh.de.

Formus Web page (2000) **Formus**, 185 Lewis Road, Suite 31, San Jose, CA 95111, USA, www.formus.com.

Fritz E (1998) Laser-sintering on its way up, **Prototyping Technology International '98**, UK & International Press, Surrey, UK, pp 186-189.

Jacobs PF (1996a) Recent Advances in Rapid Tooling from Stereolithography, White Paper, **3D Systems**, Valencia, California, USA.

Jacobs PF (1996b) Stereolithography and other RP&M technologies, **Society of Manufacturing Engineers - American Society of Mechanical Engineer**.

Klosterman DA, Chartoff RP and Pak SS (1996) Affordable, rapid composite tooling via laminated object manufacturing, **Proceedings of the International SAMPE Symposium and Exhibition**, Covina, CA, USA, Vol. 41, 1, pp 220-229.

MIT Web page (1999) **MIT,** Three Dimensional Printing Group, http://me.mit.edu/groups/tdp/.

Nakagawa T (1994) Applications of laser beam cutting to manufacturing of forming tools - Laser cut sheet laminated tool, Laser Assisted Net Shape Engineering, **Proceedings of LANE'94**, Vol. 1, pp 63 - 81

Pak SS, Klosterman DA, Priore B, Chartoff RP and Tolin DR (1997) Tooling and low volume manufacture through Laminated Object Manufacturing, **Prototyping Technology International '97**, UK & International Press, Surrey, UK, pp 184-188.

Pham DT, Dimov SS and Lacan F (1999) Selective Laser Sintering: Applications and Technological Capabilities, **Proc. IMechE, Part B: Journal of Engineering Manufacture**, Vol. 213, pp 435-449.

Pham DT, Dimov SS and Lacan F (2000) The RapidTool process: Technical Capabilities and Applications, **Proc. IMechE, Part B: Journal of Engineering Manufacture**, Vol. 214, pp 107-116.

Reinhart G and Breitinger F (1997) Rapid Tooling for Simultaneous Product and Process Development, **Proceedings of the 6th European Conference on Rapid Prototyping and Manufacturing**, Nottingham, pp 179-191.

Sachs E, Guo H, Wylonis E, Serdy J, Brancazio D, Rynerson M, Cima M and Allen S (1997) Injection molding tooling by 3D printing, **Prototyping Technology International '97**, UK & International Press, Surrey, UK, pp 322-325.

Soligen Technologies Inc. Web page (1999) **Soligen Technologies Inc.,** http://www.3dprinting.com/.

Venus AD and Van de Crommert SJ (1996) Rapid mould manufacture 'Manufacturing of injection moulds from SLS, **Proceedings of the 5th European Conference on Rapid Prototyping and Manufacturing**, Helsinki, Finland, 4-6 June, 1996, pp 171-183.

Uziel Y (1997) Seamless CAD to metal parts, **Prototyping Technology International '97**, UK & International Press, Surrey, UK, pp 234-237.

Chapter 8 Applications of Rapid Tooling Technology

The introduction of Rapid Tooling (RT) technology has enabled prototype, pre-production and in some cases full production tooling to be fabricated within significantly reduced time frames. In the previous Chapter several proprietary solutions for direct tool production were described. A sound understanding of the capabilities and limitations of these processes is essential in order to implement the technology successfully. This Chapter further discusses the DTM RapidTool process, which is one of the most significant direct tooling methods to be introduced in recent years, and from the experience gained with this process, some important lessons are highlighted, which apply to RT technology in general. Special attention is given to the specific design and finishing requirements of RapidTool inserts because these aspects have a critical effect in optimising the capabilities of the process. A study of the wear characteristics of RapidTool and EOSintM inserts is presented along with two industrial case studies which illustrate the application of RT technology in the fields of plastics injection moulding and aluminium gravity die-casting.

8.1 Insert Design

The successful utilisation of the DTM RapidTool process requires a careful analysis and understanding of design issues associated with this direct RT method. In particular, tool designers must consider process specific issues such as the type of gating, the thermal control system, the type of ejection, the type of venting, and the anticipated shrinkage [DTM, 1998]. Only after completing this analysis can the designer define parting lines and incorporate additional features into the inserts.

The RapidTool process imposes specific requirements on the insert design. SLS technology allows the cooling lines, ejector pin guides, gates, and runners to be included in the design and built directly into the inserts. This can lead to significant time savings because less effort is required for further machining of the inserts.

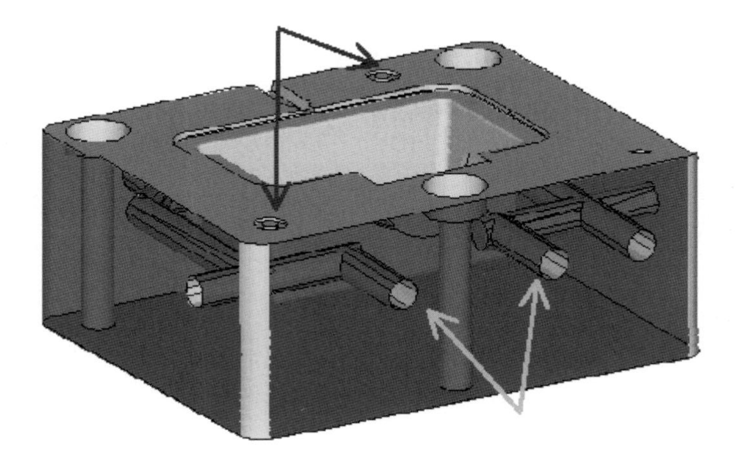

Figure 8.1 CAD model of an insert (Courtesy of DTM)

The main issues that have to be taken into account when designing RapidTool inserts are summarised below.

1. Excess material has to be added to the parting surfaces, shut-offs and sides of the inserts and machined afterwards for a good surface match. 0.23 mm of stock should be left on parting surfaces and shut-offs and 0.25 to 0.38 mm on insert exterior for further machining [DTM, 1998]. The SLS process allows material to be added to all features in the X-Y build plane by modifying the beam offset values within the build preparation software. It should be remembered, however, that adding more material than necessary could lead to a significant increase in the finishing time required.

2. CAD design of the inserts can include the cooling channels in a way that provides some degree of conformal cooling [DTM, 1998]. The cooling lines should be ∅ 6.44 mm and at least 6.4 mm below the insert surface (Figure 8.1). To facilitate the removal of the unsintered powder, cooling channels should pass through the side walls of the inserts.

3. The RapidTool process is limited to building features larger than 1mm. This is because it is very difficult to clean unsintered powder from small features at the green stage without breaking or damaging them. Also, the offset value used for the laser beam can be as high as 0.4 mm which also limits the maximum resolution achievable.

4. Features that require tight dimensional or positional tolerances but are easy to machine or can be readily added should be removed from the CAD model before the SLS process (see also Chapter 7).

5. Features such as holes across the whole part should be avoided because they can weaken it, causing the part to distort and even to collapse between the debinding and sintering stages in the furnace when its shape is maintained only by friction between the steel particles (see Chapter 7).

6. Reference datum features should be included in the CAD model to help locate insert geometry during finishing. These datum features are typically recessed in order not to change their position after machining (see Figure 8.1) [DTM, 1998].

8.2 Insert Finishing

The main difference between conventional tooling and RapidTool inserts is that in the case of conventional tooling, machining starts from a piece of material with clearly defined datums for machining and subsequent integration into a tool base. On the other hand, in the case of RapidTool inserts, there are no surfaces that can be used directly as datums and the finishing process has many similarities with the finishing of a cast part. The machining of the insert datums is critical to the final accuracy of the tools.

After copper or bronze infiltration, the surface that is least distorted by the process is the insert base plane. This is because, after debinding in the furnace cycle, the steel powder is self-supporting and, under its own weight, the part tends to sag and copy the shape of the alumina plate underneath. As a result, some of the distortions introduced during the cross-linking step can be corrected. This base plane has to be machined first and then used as the primary datum plane for the next machining steps.

The second plane to be employed as a datum plane is a side plane opposite to the plane through which the part is infiltrated. This plane is free of infiltration marks and is perpendicular to the first datum plane (insert base plane). Once machined, it can be used as secondary datum for the next machining steps.

Either of the other two remaining side walls of the insert can be adopted as a tertiary datum plane.

Because of the stair effect on slopes, finishing and polishing of a RapidTool insert can be a laborious process. The finishing and polishing efforts can be significantly reduced by initial finishing of the inserts prior to infiltration. Fine files and polishing stones can remove 0.075 to 0.10 mm of material to limit the stair step effect.

After inserts are infiltrated to achieve the required accuracy and surface finish the following machining and finishing operations can be employed [DTM, 1998a; DTM, 1998b; DTM, 1998c]:

1. *Milling using high-speed or carbide cutting tools*. Carbide tools are more effective because of a hard scale layer on the infiltrated parts.

2. *Welding/Brazing*. Due to the material being a matrix of dissimilar metals, RapidTool inserts are very difficult to weld. However, particular brazing rods can be used to join and repair RapidTool inserts.

3. *EDM (Electro Discharge Machining)*. Wire and volume EDM equipment can be employed to machine RapidTool inserts. For volume EDM, the same process parameters as those used for tool steel can be applied. For wire EDM, the process settings for cutting tool steel have to be modified to reflect the specific material properties of RapidSteel.

4. *Chemical Etching and Plating*. Chemical etching can be used to achieve a texturing effect on infiltrated parts. RapidTool inserts can be also plated with chrome, nickel and electroless nickel.

8.3 Rapid Tooling Inserts Wear Resistance

Injection moulding tools normally consist of many parts in motion relative to one another and to the plastic material. Abrasion, adhesion, corrosion and fatigue are common problems, and several types of wear can be observed at the same time [DTI, 2000].

To date, no test specific to the study of mould wear has yet been developed. In order to study the wear process in an injection mould and how this could affect the life of rapid tooling inserts, a test has been specifically designed in the authors' laboratory [Lacan, 2000] to monitor the degradation of the two types of commercially available Rapid Tooling processes: DTM RapidTool and EOSINT M.

Although both processes involve selective laser sintering, the RapidTool and EOSintM processes are very different and so are the properties of the materials used and the parts produced. One common feature of the two processes is that the final parts are both made of more than one material, a hard material and other softer materials.

As described in Chapter 7, RapidTool parts are composed of sintered stainless steel particles infiltrated with bronze. On the other hand, EOSINT M parts are made of nickel powder liquid-sintered to a low melting bronze alloy and finally infiltrated with an epoxy resin. Although there has been no study reporting in detail the wear mechanism of SLS metal mould inserts, the softest material is in both cases likely to be the first to deteriorate during plastic injection moulding. Because SLS metal inserts are made of at least two different materials, conventional surface heat treatment cannot be used to harden the surface and coatings have to be applied to prevent premature wear.

Coatings that act as barriers between the erosive/abrasive particles and the mould surface can protect it but the selection of a suitable coating has so far been difficult. Predicting erosion/abrasion performances for RapidTool insert coatings is more complex than for bulk material. Even for the latter, erosion cannot be predicted accurately as properties that are difficult to determine, such as thermal expansion, bond strength, residual stresses and thickness, need to be considered. The usual method of handling erosion/abrasion problems is to replace the eroded surface with a harder one [Wood, 1999].

The paper clip depicted in Figure 8.2 was selected as the moulding for testing the wear properties of rapid tooling inserts for several reasons.

- A large number of mouldings were expected before the mould showed any sign of wear. The part produced had to have small dimensions in order to limit the amount of plastic needed for the tests.

- The paper clip was designed so that only one mould cavity was required. This halved the number of inserts to be produced and considerably reduced the machining time needed for integrating them.

- Due to their small size (40x50x12mm), the paper clip inserts could readily be mounted onto the available equipment for surface inspection.

As seen in Figure 8.2, the design of the paper clip incorporates a square feature in the centre of the part. This feature was specially added to monitor the surface wear.

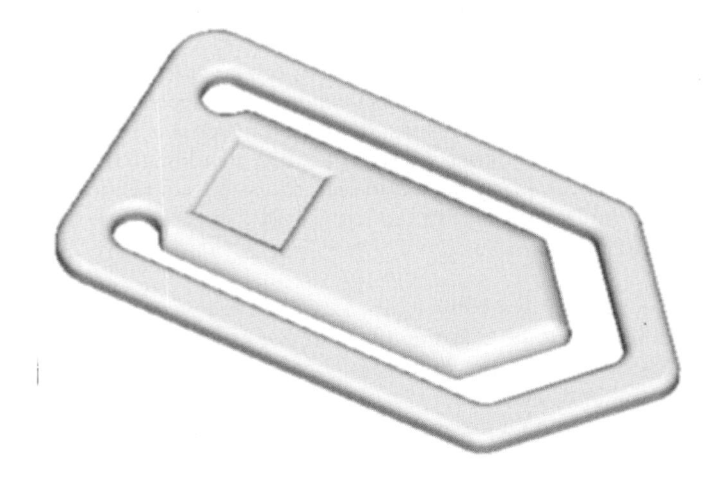

Figure 8.2 Paper clip design (40x20x1.5mm)

Six RT and two aluminium inserts were built and assembled into a multi-cavity tool. The two aluminium inserts were included in the test for comparison purposes. Four of the RT inserts were plasma-spray coated. Plasma spray coating was investigated because it is quick to apply and a wide range of coating materials, including those with high melting points, is available. Two materials were chosen for their wear resistance: stellite and molybdenum (COHFANF6 and HCST-Amperit from HC Starch Corp.). A listing of the RT inserts and their coatings used in the wear tests is given in Figure 8.3. To ensure the same filling conditions (pressure, temperature and flow), the inserts had to be located symmetrically and have gates and runners with the same dimensions. The moulding conditions and material are as follows:

- Moulding material: PA6 30% glass filled
- Processing conditions:
  - Melt temperature: 275 °C
  - Mould temperature: 60 °C
  - Maximum injection pressure: 60 MPa
  - Injection time: 0.7 s

In order to accelerate the wear rate, 30% glass-filled nylon was selected as the moulding material.

Insert material	Coating	Coating thickness (μm)	Roughness, Ra (μm)
Rapid Steel 2.0	no coating	N/A	0.1
EOSint M	no coating	N/A	2.2
Rapid Steel 2.0	Molybdenum	110	5.7
EOSint M	Molybdenum	30	6
Rapid Steel 2.0	Stellite	90	4.8
EOSint M	Stellite	50	5.1

Figure 8.3 Wear test inserts and coatings

8.3.1 Wear Test Results

8.3.1.1 Non-coated RapidSteel 2.0 Insert

To save time for the finishing and integration of the inserts into the mould base and because the paper clip shape was very simple, the RapidSteel 2.0 inserts used in the tests were machined directly from a piece of material previously produced by the RapidTool process. As a result, the surface finish of the inserts is very good.

The non-coated insert proved to be able to withstand the injection of glass filled nylon very well. Visually, the insert appeared completely unchanged and, after 16000 shots, surface profiler measurements taken with a pitch of 0.1 mm revealed no sign of wear. Microscope examinations showed only very slight changes at the 500x magnification used throughout the tests.

Photographs of the tested insert and of a spare insert were also taken at the higher magnification of 2000x (Figures 8.4 and 8.5). They show differences for the square feature and the gate that are not visible in the scanning results. To confirm these changes, the square feature and the gate were scanned with a profiler using this time a smaller pitch of 10 μm. The results contradict the visual inspection and show no significant differences for the roughness and the waviness values.

Figure 8.4 Micrographs of the square feature surface of the non-coated RapidSteel 2.0 insert after 0 and 16000 injections (magnification 2000x)

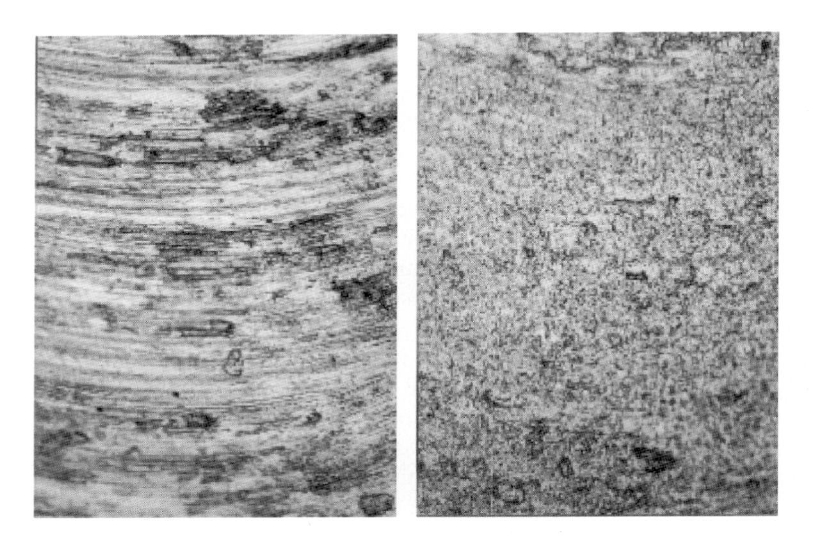

Figure 8.5 Micrographs of the gate surface of the non-coated RapidSteel 2.0 insert after 0 and 16000 injections (magnification 2000x)

A probable explanation for this phenomenon is that because bronze is much softer than steel, when the part was machined, steel was cut but some bronze was spread onto the part surface. This excess bronze seen under the microscope was subsequently removed during the wear tests.

8.3.1.2 Non-coated EOSINT M Insert

The EOSINT M inserts were produced by EOS for this study. They were machined to size for integration with the bolster, but no additional finishing was performed on their surfaces to improve the finish. As a result, the roughness of the EOSintM inserts is higher than for the RapidSteel 2.0 inserts.

The non-coated EOSINT M insert is by far the softest of all the inserts tested. As a result, a much quicker wear rate was expected.

Measurements of the surface of the square feature revealed no significant change in the roughness and waviness. The slight differences noted are probably due to differences in the location of the insert in the profiler. Profiles taken of similar areas of the square surface show very similar shapes and roughness values during the tests.

Although the colour of the mould cavity changed slightly during the tests, microscopic observations of the square surface did not show up any major sign of wear. A micrograph of the surface of the square feature can be seen in Figure 8.6.

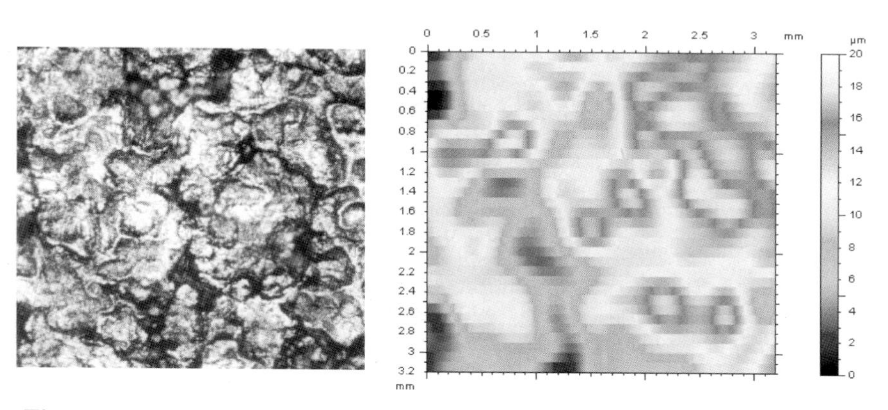

Figure 8.6* Micrograph and waviness colour coded map of the square feature surface of the EOSINT M insert after 16000 injections (magnification 1000x)

The results are different for the gate where the higher plastic flow velocity must have caused more damage. After 2500 injections, small particles of material began to be torn off the gate surface. Figure 8.7 shows the state of the gate surface at different stages during the tests. Degradations are hardly visible to the naked eye and too small to have any effect on the moulded part. An estimate of the amount of material removed during the tests can be obtained using the Talymap software. For example, Figure 8.8 shows the depth of the holes left by the removed particles and the volume of the hole in the encircled area in Figure 8.9 is approximately $0.26mm^3$.

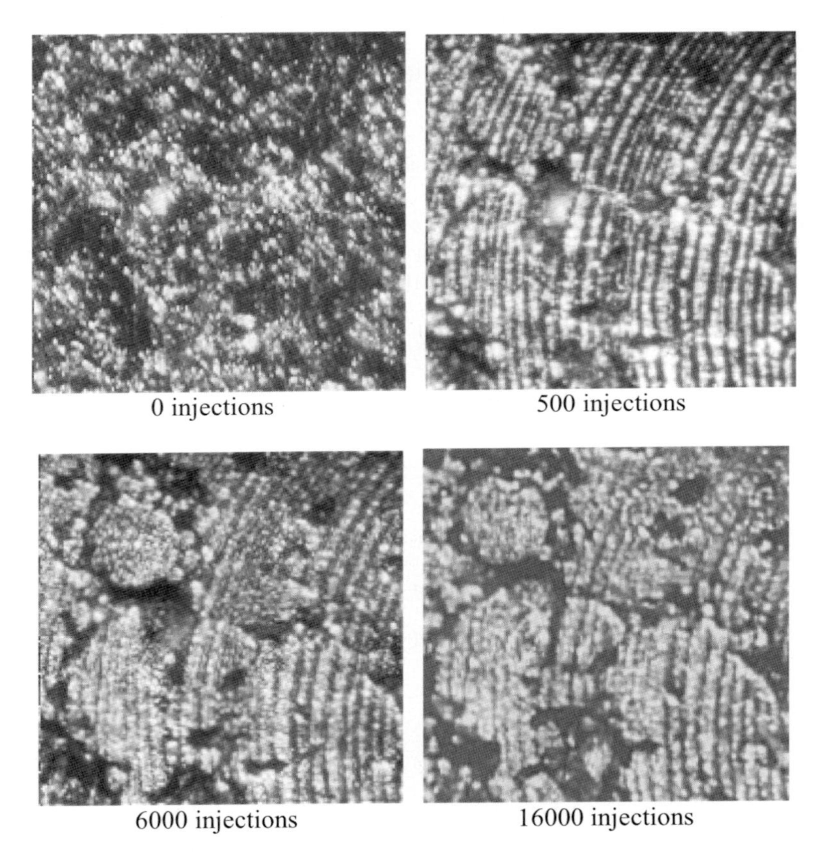

Figure 8.7 Changes in the gate surface for the non-coated EOSINT M insert (magnification 500x)

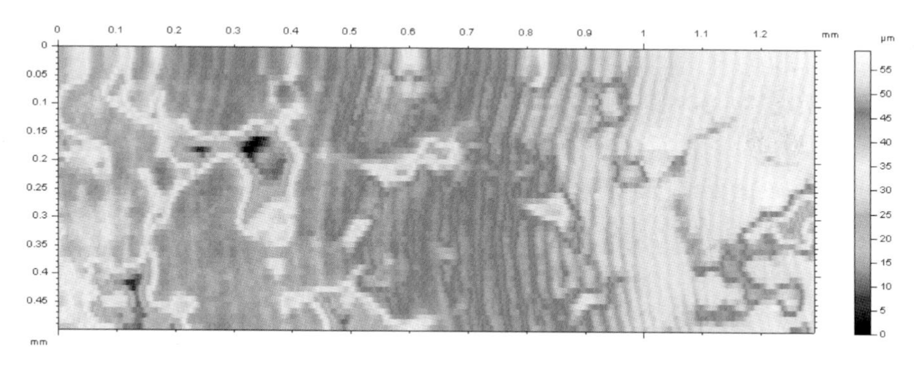

Figure 8.8* Colour coded depth map and micrograph (magnification 1000x) of the gate of the non-coated EOSintM insert after 16000 injections

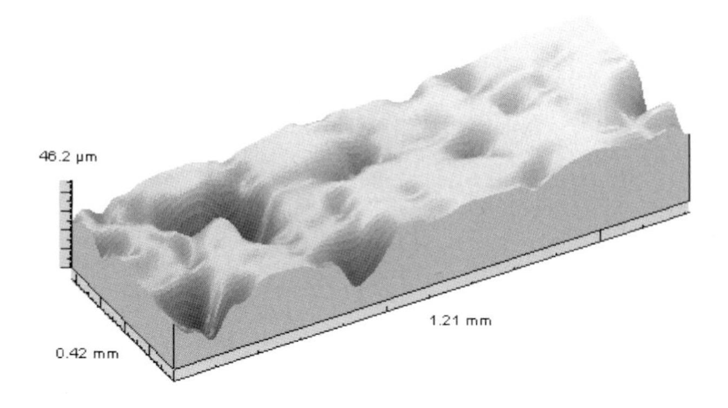

Figure 8.9 3D map of the gate of the non-coated EOSINT M insert after 16000 injections (roughness removed)

8.3.1.3 Spray-coated Inserts

The two plasma spray-coatings used withstood the injection moulding trials remarkably well.

The micrographs show no noticeable changes in the surfaces of the inserts and the profiler results indicate no sign of wear either. The only few observable differences are again probably due to the fact that the part could not be scanned twice at exactly the same location.

Figures 8.10 and 8.11 show the surface of the square feature and its waviness for the four spray-coated inserts. By comparing the waviness values from these figures with those obtained for the non-coated inserts it can be seen that the application of the coating worsened the surface finish.

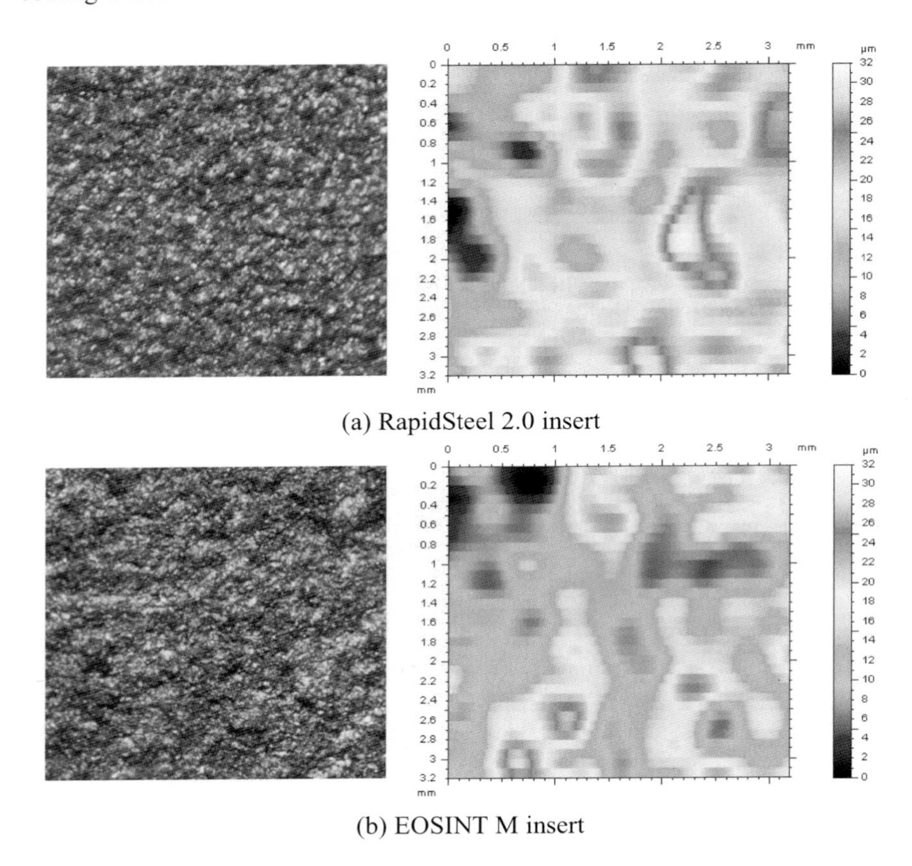

(a) RapidSteel 2.0 insert

(b) EOSINT M insert

Figure 8.10* Surface (magnification 250x) and waviness of the square feature for the Molybdenum-coated RapidSteel 2.0 and EOSINT M inserts after 8000 injections

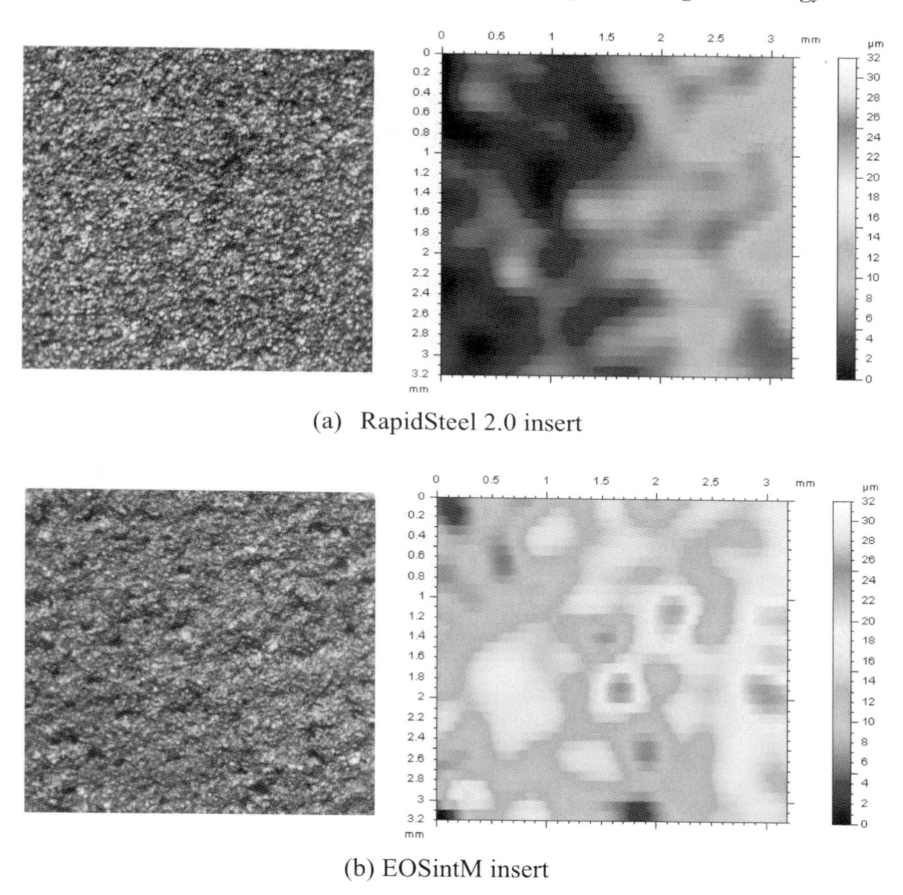

(a) RapidSteel 2.0 insert

(b) EOSintM insert

Figure 8.11* Surface (magnification 250x) and waviness of the square feature for the Stellite-coated RapidSteel 2.0 and EOSINT M inserts after 8000 injections

Note that the magnification had to be kept below 500x due to the high surface roughness of the sprayed parts.

Unfortunately, the test conditions were not harsh enough to compare the behaviours of the two kinds of coating as neither the micrographs nor the profile measurements could reveal any difference before, during or after the tests for the different coated surfaces, even in the area of the gate.

8.3.2 Discussion of the Wear Test Results

The results of the wear tests show good wear resistance for most of the inserts tested.

The EOSINT M inserts performed the worst. However, apart from the gates which were severely damaged, the rest of the parts showed no measurable wear. This suggests that the life of EOSINT M moulds can be greatly improved by adding a conventionally manufactured steel gate to the insert.

The RapidSteel 2.0 inserts showed no measurable sign of wear on the whole part. The micrographs suggest that some bronze on the part surface mainly in the area of the gate might have been brushed away, but results indicate that a much higher number of parts can be expected from the inserts before any measurable wear appears on the tool surface.

The two plasma spray-coatings tested yielded better results than expected. There was no problem of adhesion between them and the two SLS substrates as they show absolutely no sign of delamination or deterioration in the microsurface observations and the profile measurements. However, it can be noted that the surface finish of the sprayed parts worsened after the coating deposition and that, although the RapidSteel 2.0 and EOSintM parts had been coated at the same time, the final coating thicknesses were different for the two types of inserts. Further research needs to be carried out to address these problems.

8.4 Case Studies

8.4.1 ABS Portable Electronic Tour Guide

An initial batch of 60 sets of front and back covers of a portable electronic tour guide (Figure 8.12) used in places such as museums, had to be manufactured in four weeks. Because of this short lead-time, it was decided to build mould inserts using the RapidTool process.

The mould inserts were designed in a 3D CAD package and two pairs of inserts were built, one for the front cover and one for the back cover. Each pair of inserts took five days to produce using RapidSteel 1.0. Due to time pressure, the mould design was simplified and the mould was produced without sliding inserts and cooling lines for the initial batch (Figure 8.13).

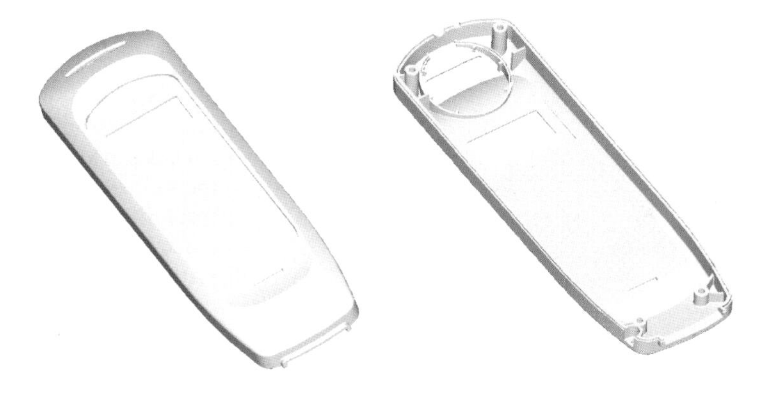

Figure 8.12 3D CAD representation of the tourguide front cover

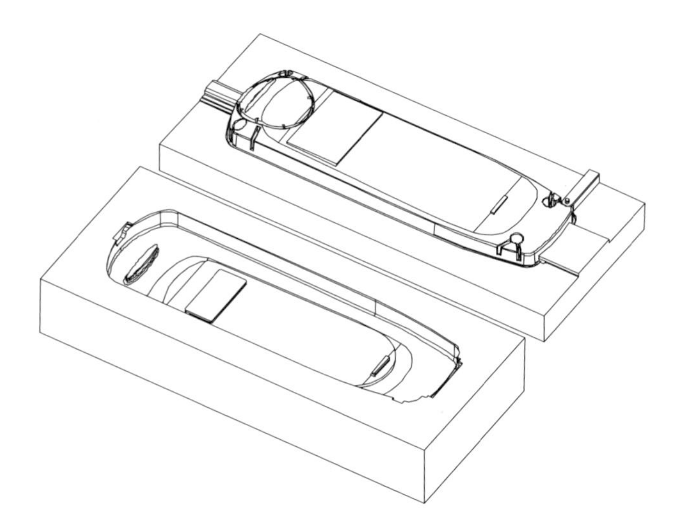

Figure 8.13 Design of the tourguide front cover injection moulding inserts

The inserts were finished following the steps outlined in Figure 8.14. Polishing of the inserts took slightly longer than for conventional steel inserts because of the stair stepping on some surfaces.

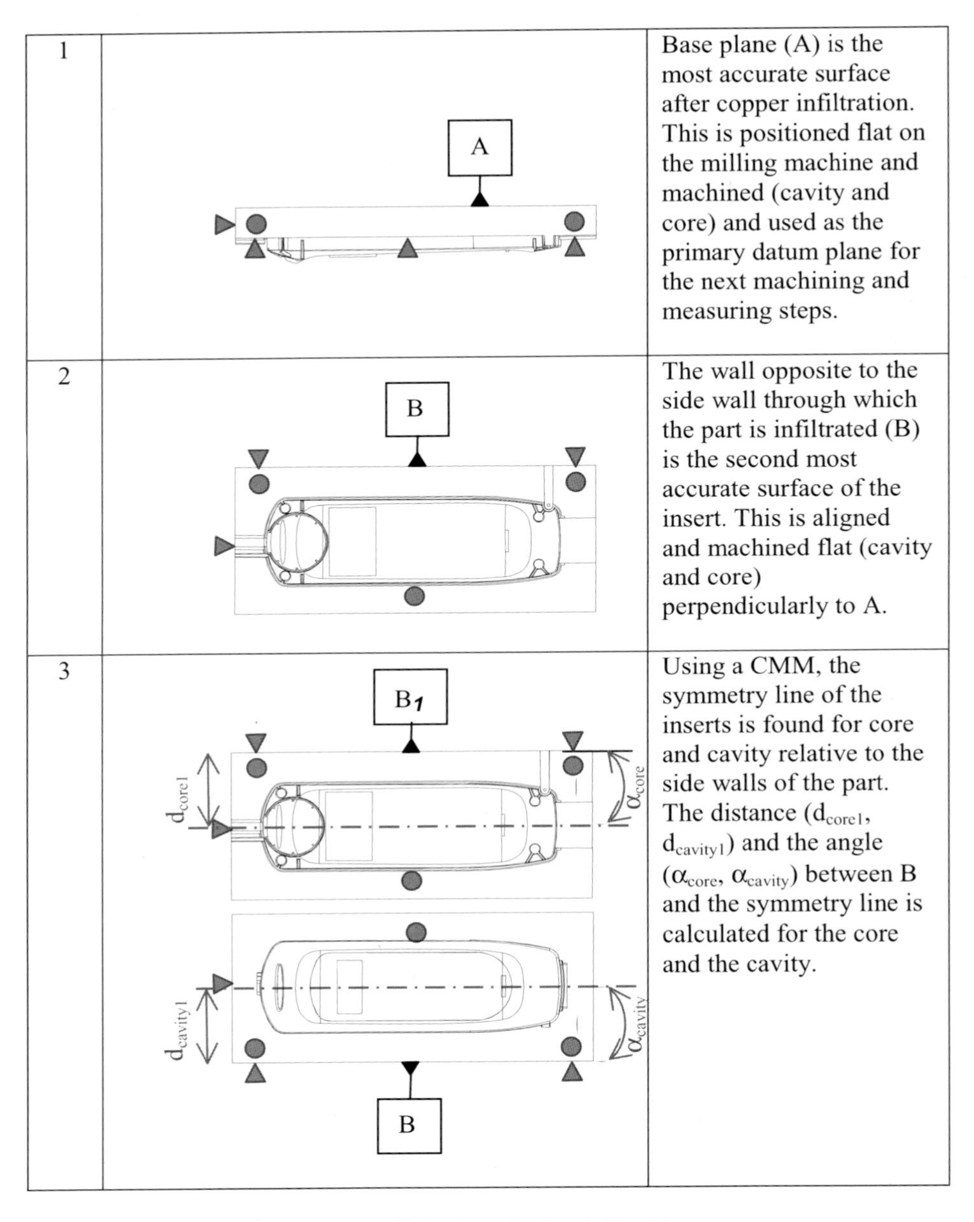

Step	Description
1	Base plane (A) is the most accurate surface after copper infiltration. This is positioned flat on the milling machine and machined (cavity and core) and used as the primary datum plane for the next machining and measuring steps.
2	The wall opposite to the side wall through which the part is infiltrated (B) is the second most accurate surface of the insert. This is aligned and machined flat (cavity and core) perpendicularly to A.
3	Using a CMM, the symmetry line of the inserts is found for core and cavity relative to the side walls of the part. The distance (d_{core1}, $d_{cavity1}$) and the angle (α_{core}, α_{cavity}) between B and the symmetry line is calculated for the core and the cavity.

Figure 8.14 Finishing of a RapidTool insert

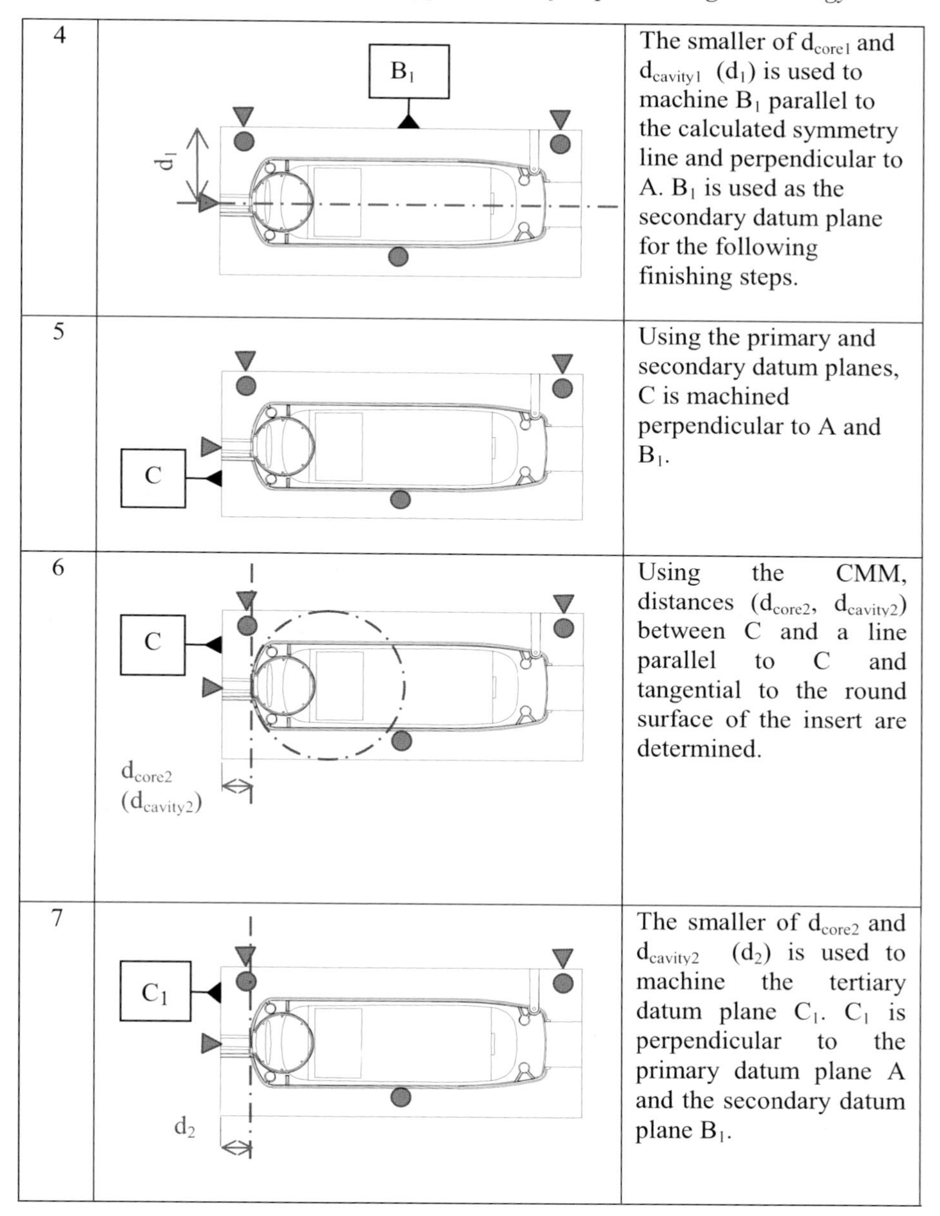

4		The smaller of d_{core1} and $d_{cavity1}$ (d_1) is used to machine B_1 parallel to the calculated symmetry line and perpendicular to A. B_1 is used as the secondary datum plane for the following finishing steps.
5		Using the primary and secondary datum planes, C is machined perpendicular to A and B_1.
6		Using the CMM, distances (d_{core2}, $d_{cavity2}$) between C and a line parallel to C and tangential to the round surface of the insert are determined.
7		The smaller of d_{core2} and $d_{cavity2}$ (d_2) is used to machine the tertiary datum plane C_1. C_1 is perpendicular to the primary datum plane A and the secondary datum plane B_1.

Figure 8.14 Finishing of a RapidTool insert (continued)

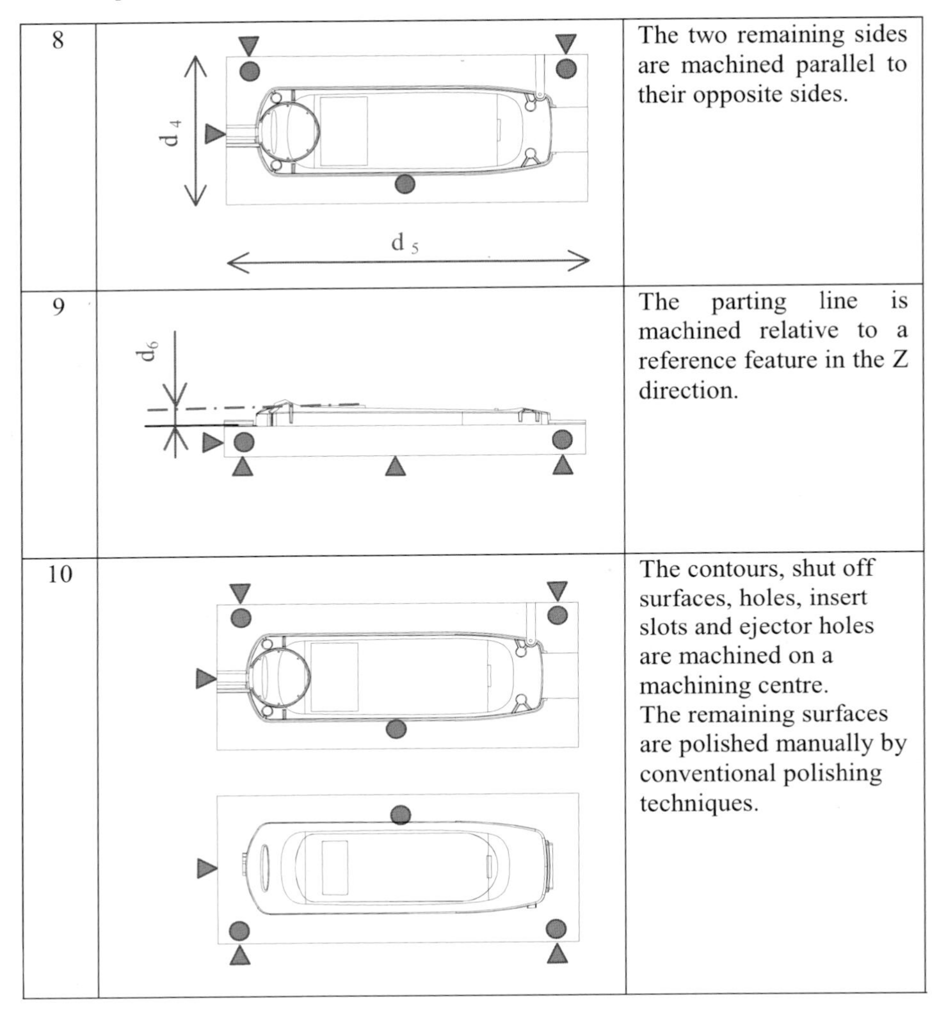

Step	Description
8	The two remaining sides are machined parallel to their opposite sides.
9	The parting line is machined relative to a reference feature in the Z direction.
10	The contours, shut off surfaces, holes, insert slots and ejector holes are machined on a machining centre. The remaining surfaces are polished manually by conventional polishing techniques.

Figure 8.14 Finishing of a RapidTool insert (continued)

A local company integrated the inserts into the mould base (Figure 8.15) and moulded the parts. Given the thermal conductivity of the inserts, the cooling channels were not necessary in this case because only a small number of parts were to be produced each time.

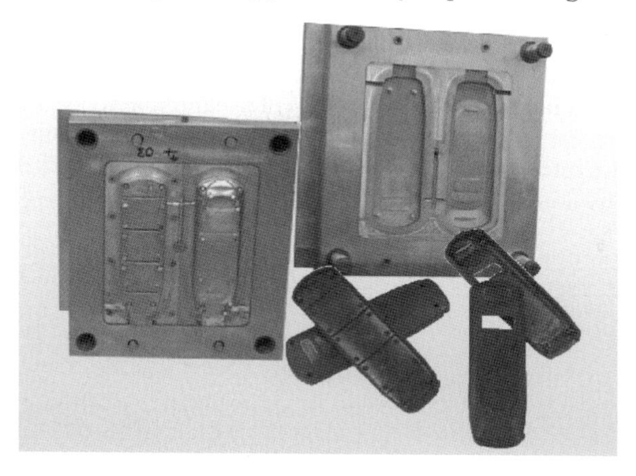

Figure 8.15 Finished inserts

The accuracy of injection moulded parts depends greatly on the quality of finishing and polishing. To obtain an idea of the accuracy achievable under real processing conditions, the external surface of a moulded front cover was scanned on a co-ordinate measuring machine. Using a module within the CAD package for verifying part dimensions, the measured moulding was compared to the 3D CAD model. The result of the comparison is given in Figure 8.16. From the measurements, 68.5% of the dimensions were within ± 0.15 mm and 90% within - 0.25 and +0.5 mm. There were deviations of over 0.5 mm from the nominal dimensions but they appeared on sharp edges that were finished manually.

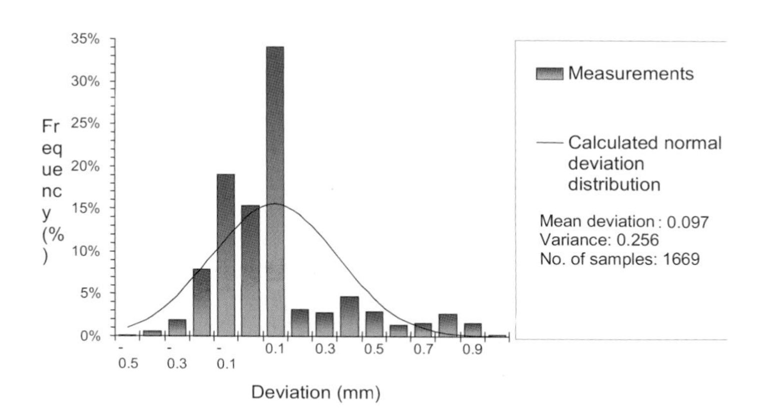

Figure 8.16 Comparison between the front moulding and its 3D CAD model (deviation distribution)

8.4.2 Aluminium Windscreen Wiper Arm

To evaluate the suitability of RapidTool for aluminium gravity die-casting, RapidTool inserts for a windscreen wiper arm were built using RapidSteel 1.0 (Figure 8.17). The integration of the inserts into a casting rig and the trial castings were carried out by a local company.

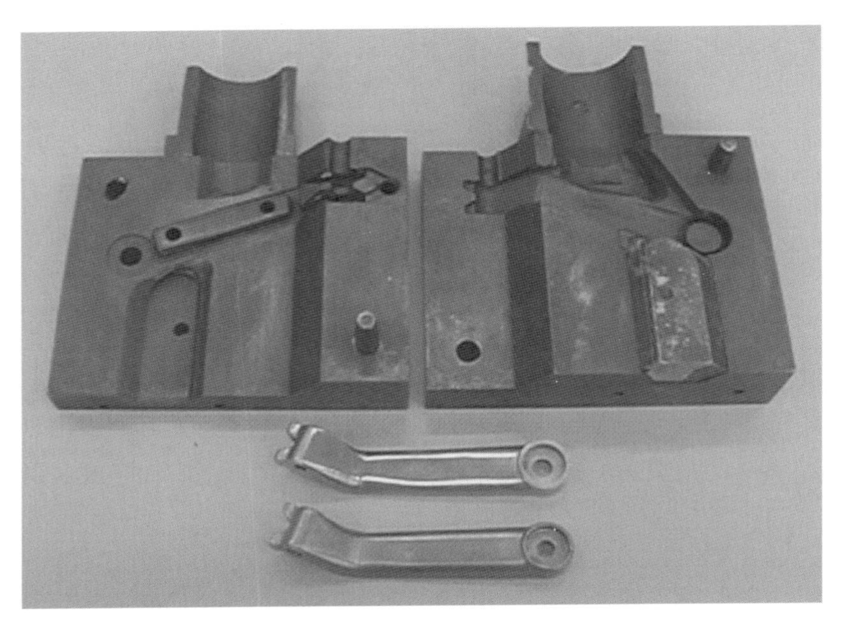

Figure 8.17 Windscreen wiper arm die-casting inserts and parts

The inserts were finished following the steps described in Figure 8.14. One of the bosses at the end of the wiper was used as a reference feature to achieve good matching of the two halves of the tool. The machining of the parting line proved more difficult than for the tour guide inserts. The reason for this was the more complex shape of the parting line which lies in three different planes.

The aluminium alloy used was LM6. The original part was designed for pressure die-casting and this accounted for initial problems with metal flow to the thin sections of the die. This problem was exacerbated by the high thermal conductivity of RapidSteel 1.0 which exceeds that of cast iron: the castings solidified very quickly so that the metal had to be introduced rapidly. To evaluate the variations of temperature in the die-casting tool a thermocouple was positioned 5mm from the surface of the core and the cavity (Figure 8.18). The results show that the die reached the casting temperature after only fifteen castings and the casting cycle time was about 3 minutes. This confirms the excellent thermal properties of RapidSteel which allows shorter cycle times. However, this needs to be taken into consideration when designing runner systems as the heat loss in the die is significant. In the present case, the runner system was repositioned, putting the thinnest casting section at the lowest possible point. This increased the head above that section to facilitate the flow of metal to it.

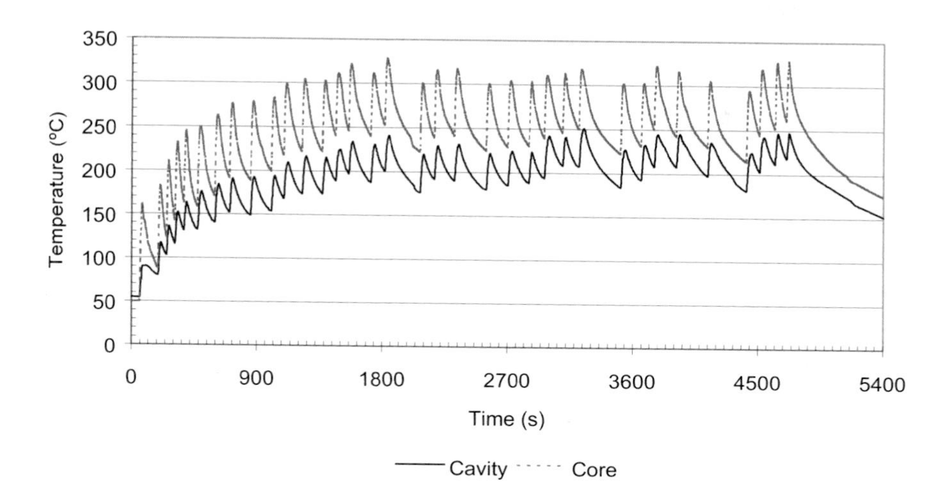

Figure 8.18 Variations of the temperatures of the core and cavity die-casting inserts

Another problem was that the cast part proved difficult to eject from the die because the die blocks had not been polished other than to clean up the mating faces. To overcome this, additional ejectors and careful polishing of the cavity were necessary in critical areas to prevent the casting from catching and distorting at the ejection stage.

Some reworking was carried out on the dies as these problems were discovered, but the main problems were caused by the complexity of the part geometry rather than the RapidTool process itself. The final definition of the casting was good and, after 250 castings in four separate runs, no signs of degradation were visible either on the insert surfaces or on the cast parts.

The tests showed that RapidSteel 1.0 dies can be utilised for the production of low to medium size batches of castings. Given the quality of the die material, it is estimated that over five thousand castings could easily be produced from the dies.

8.5 Summary

This Chapter has examined three practical issues associated with the application of the RapidTool process to the production of tooling inserts: insert design, insert finishing and insert wear resistance. The process has limitations which must be studied and understood before applying it for tool production. The experience of different users [DTM, 1998b; Pham et al., 1999 and 2000] has demonstrated that the process can be successfully utilised for manufacture of prototype and pre-production tools for injection moulding and die-casting.

With the introduction of new materials, it is expected that higher accuracy will be achieved. The use of smaller steel particles and thinner layers will mean further improvements in accuracy due to the reduction of stair stepping. In addition, processing speeds will increase because fewer stages are involved and less finishing effort will be needed to remove stair stepping effects. Thus, the RapidTool process can be considered to have evolved into a more mature and reliable technology that should in the near future enable rapid tooling to become more widely accepted in the moulding and casting industries.

References

DTI Web page (2000) **Danish Technological Institute,** Tribology Centre, Aarhus, Denmark, http://www.tribology.dti.dk/coatings.html.

DTM Press Release (1998a) RapidSteel 2.0® Reduces Typical Build Times by Half, Improves Surface Finish in Applications for Metal Mold Inserts, **DTM Corporation**, http://www.dtm-corp.com/News/newsdesk.htm#Press.

DTM RapidTooling Users Group Meeting Report (1998b) **DTM Corporation**, 1611 Headway Circle, Building 2, Austin, TX.

DTM Product information (1998c) RapidSteel 2.0 Mold Inserts for Plastic Injection Molding, **DTM Corporation**, 81611 Headway Circle, Building 2, Austin, TX.

Lacan F. (2000) Capabilities of the RapidTool process, **PhD Thesis**, Cardiff University, UK.

Pham DT, Dimov SS and Lacan F (1999) Selective Laser Sintering: Applications and Technological Capabilities, **Proc. IMechE, Part B: Journal of Engineering Manufacture**, Vol. 213, pp 435-449.

Pham DT, Dimov SS and Lacan F (2000) The RapidTool Process: Technical Capabilities and Applications, **Proc. IMechE, Part B: Journal of Engineering Manufacture**, Vol. 214, pp 107-116.

Wood RJK (1999) The Sand Erosion Perfomance of Coatings, **Materials and Design**, Vol. 20, 4, pp 179-191.

Chapter 9 Rapid Prototyping Process Optimisation

Initially conceived for design approval and part verification, RP now meets the needs of a much wider range of applications. In order to satisfy the requirements of these applications and broaden the area of application of RP, significant research and technology development is being directed towards process optimisation.

In Chapters 4 and 5, together with the description of commercially available RP systems, the latest developments in RP materials and process-specific build techniques have been discussed. This chapter focuses on other aspects of RP process optimisation such as analysis of factors influencing the accuracy of RP processes and selection of optimal part build direction.

9.1 Factors Influencing Accuracy

The accuracy of a RP process is difficult to predict as it is a function of many different factors, some of which can be interdependent. The factors that most influence RP process accuracy can be considered in three groups [Pham and Ji, 2000]. The first group includes factors causing errors during the data preparation stage such as STL file generation, model slicing and part build direction. The second group includes factors influencing the part accuracy during the build stage such as process specific parameters. The third group of factors is directly related to the part finishing techniques employed.

9.1.1 Data Preparation

9.1.1.1 Errors due to Tessellation

Most RP systems employ standard STL input files. A STL file approximates the surface of the 3D CAD model by triangles. Errors caused by tessellation are usually ignored because of the belief that tessellation errors can be minimised by increasing the number of triangles. However, in practice the number of triangles cannot be increased indefinitely. The resolution of STL files can be controlled during their generation in a 3D CAD system through tessellation parameters. For example in Pro/Engineer, the STL generation process can be controlled by specifying the chord height or the angle control factor [Williams et al., 1996]. In particular:

- *Chord Height.* This parameter specifies the maximum distance between a chord and surface (Figure 9.1). If less deviation from the actual part surface is required, a smaller chord height should be specified. The lower bound for this parameter is a function of the CAD model accuracy. The upper bound depends on the model size.

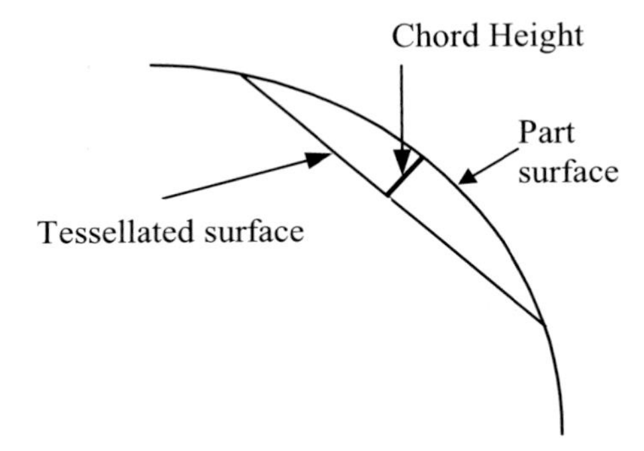

Figure 9.1 Chord height

- *Angle Control.* This parameter specifies the required definition level along curves with small radius. Specifically, it defines a threshold for the curve radius (r_0) below which the curves should be tessellated:

$$r < r_0 = \frac{partsize}{10}$$

to achieve a maximum chord height of:

$$\left(\frac{r}{r_0}\right)^{\alpha} \textit{Chord Height}$$

where *partsize* is defined as the diagonal of an imaginary box drawn around the part and α is the Angle Control value. Specifying a nil value for α will mean no additional improvement for curves with small radii.

To achieve a better part accuracy, tessellation errors have to be taken into account. For example, if the part is large, a feature with a small radius will be tessellated poorly. Suppose a model with overall dimensions of 250x250x250 mm has a round corner with a radius of 1 mm. The results of tessellating the model by applying Chord Heights of 0.5 and 0.05 mm respectively are shown in Figure 9.2. Unfortunately, the increase of Chord Height leads not only to smoother surfaces but also to larger data files. Therefore, a compromise parameter value should be selected to obtain the best trade-off between accuracy and file size.

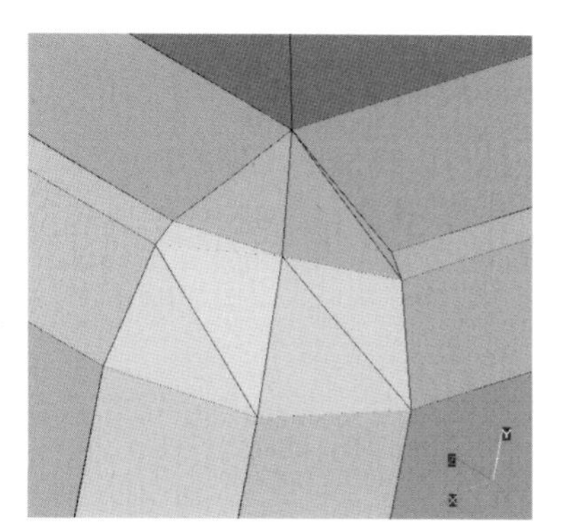

Figure 9.2 STL files generated by applying Chord Heights of 0.5 mm (left) and 0.05 mm (right)

9.1.1.2 Errors due to Slicing

RP processes have a stair-stepping problem that is found in all layer manufacturing technologies. Stair-stepping is a consequence of the addition of material in layers. As a result of this discrete layering, the shape of the original CAD models in the build direction (z) is approximated with stair-steps. This type of error is due to the working principles of RP processes, which can be assessed in data preparation.

Mathematically, curves are described with curvatures and a curvature radius. In engineering, a curve can be replaced with arcs that have common tangent lines, curvatures and concavity directions at the same point. Similarly, curves in a section of a CAD model can be replaced with arcs. To assess the error of stair-steps, arcs can also be used.

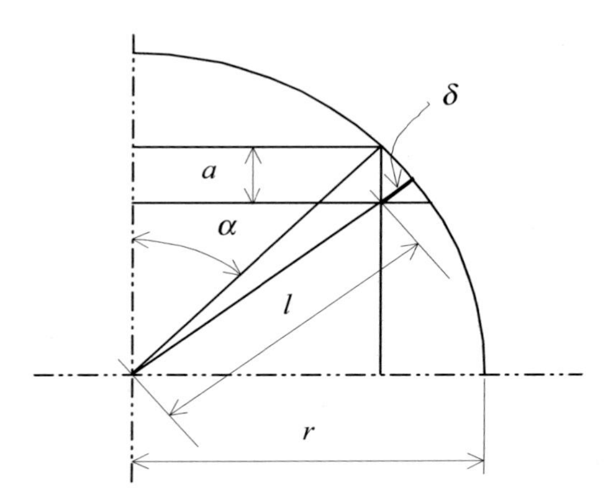

Figure 9.3 Error due to replacement of arcs with stair-steps

The error due to the replacement of a circular arc with stair-steps is illustrated in Figure 9.3 and defined as

$$\delta = r - l \tag{9.1}$$

where δ is the cusp height, r the radius and l the difference between r and δ. When l is at its minimum value, δ will reach its maximum value. Hence

$$\delta_{\max} = r - l_{\min} \tag{9.2}$$

From Figure 9.3, l is given by

$$l = \sqrt{r^2 + a^2 - 2ar\cos\alpha} \tag{9.3}$$

where a is the layer thickness, r the radius and α the angle subtended by the top portion of the arc. When $\alpha = 0$,

$$l_{\min} = r - a \tag{9.4}$$

Substituting equation 9.4 for l_{min} in equation 9.2 gives

$$\delta_{\max} = a \tag{9.5}$$

The above analysis indicates that the maximum error due to the replacement of an arc with stair-steps, the cusp height δ_{max}, which is equal to the layer thickness, occurs at the top of the arc where the tangent line is horizontal. For other general curves, the maximum cusp height will occur at points where the tangent lines are nearly horizontal.

The stair-steps particularly affect slight slopes [Reeves et al., 1996]. This problem influences mainly the roughness of the part and can be alleviated by reducing the thickness of the layers. However, layer thickness cannot be indefinitely decreased and a compromise has to be found between thickness and build speed. This problem can be partially overcome using adaptive slicing which generates different slice thicknesses based on the local slope of the part (Figure 9.4) [Rause et al., 1997].

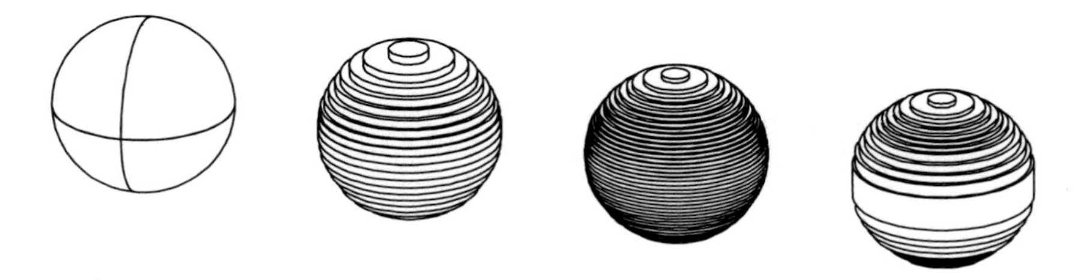

Figure 9.4 Slicing of a ball (A: No slicing, B: Thick slicing, C: Thin slicing, D: Adaptive slicing)

There are two types of errors resulting from slicing. One is because of mismatching in height between slice positions and feature boundaries; the other is the replacement of polygons with stair-steps.

Figure 9.5 illustrates the mismatching effect. The broken lines are slicing positions, which do not pass exactly through the bottom point and top point on the circle. The top side and bottom side of the feature will be built as the shaded layers in the figure. Thus, the mismatch error can be defined as

$$\delta_{mm} = \delta_{tm} + \delta_{bm} \tag{9.6}$$

where δ_{tm} is the mismatch error at the top and δ_{bm} the mismatch error at the bottom. The mismatch error δ_{mm} can be as much as two layers.

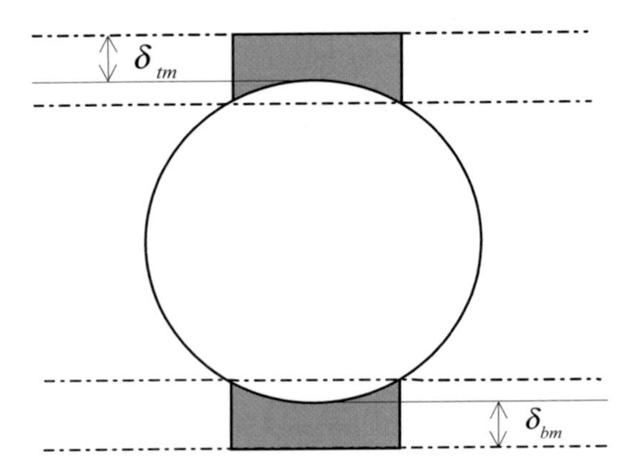

Figure 9.5 Mismatching in height

Figure 9.6 illustrates the error resulting from slicing tessellated arcs. The broken lines represent the stair-steps when the arc is directly sliced. The corresponding solid lines represent the stair-steps formed by slicing the chord (tessellated arc). Thus the slicing error can be defined as

$$\delta_{st} = \delta_s + h_{chord} \tag{9.7}$$

where $\delta_s = a.\cos\alpha$, α is the angle between the chord and the horizontal axis, and h_{chord} is the chord height. When slicing a STL file, the error consists of the tessellation error h_{chord} and the cusp height error δ_s. Thus, the error is larger than that resulting from directly slicing the original CAD model. Also the maximum errors happen with chords that have the smallest values of α, which is similar to the case when the CAD model is directly sliced.

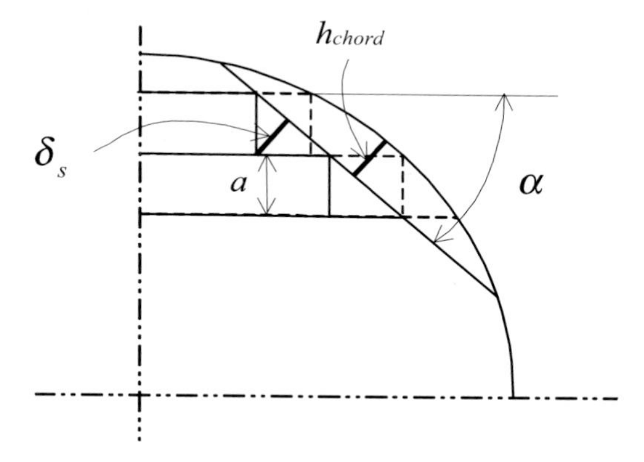

Figure 9.6 Step error due to slicing tessellated arcs.

9.1.2 Part Building

Different process-specific errors occur during part building. For example, material shrinkage and control errors are the most significant factors influencing the part accuracy when thermal RP processes are employed, e.g. SLS, MJM, FDM, LENS, etc. On the other hand, curing and control errors are the main factors in RP processes such as SL and SGC. In this section, specific part building errors resulting from SL and SLS processes are discussed.

9.1.2.1 Part Building Errors in the SL Process

There are two main types of errors in the part building process, namely curing errors and control errors. Curing errors refer to those errors that are caused by over-curing and scanned line shape. Control errors are those errors caused by layer thickness and scan position control. Both types of errors affect part accuracy.

- *Over-curing.* Laser over-curing is necessary to adhere layers to form solid parts. However, it causes dimensional and positional errors to features. The over-cured material in the bottom layer can be seen in Figure 9.7a, from which the unusual thickness of the bottom layer can be noted. This causes a dimensional change in the z direction along the lower feature boundary. As a result, the feature shape is deformed and the feature centre position is shifted. Figure 9.7b shows part of a deformed boundary of what should have been a circular feature.

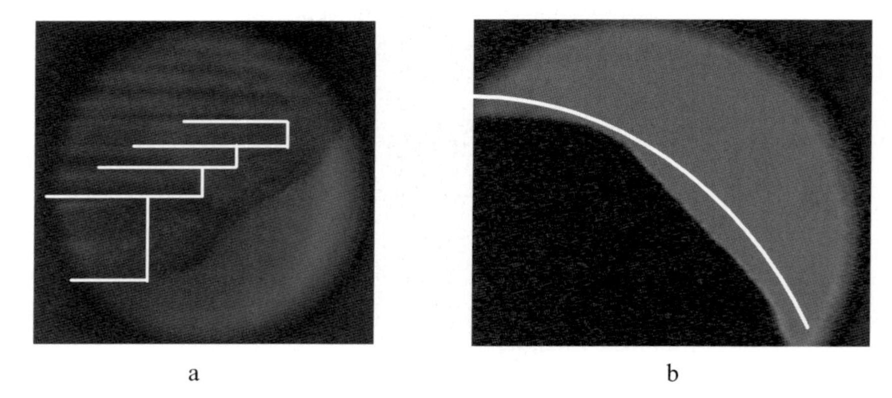

Figure 9.7* Over-curing effects on accuracy (a - thicker bottom layer, b - deformed hole boundary)

- *Scanned line shape.* A scanned line is created when a laser beam scans the resin surface. The cross section of the scanned line is referred to as the scanned line shape. A cross section is shown in Figure 9.8. Note that the shape deviates from the theoretically predicted parabola [Jacobs, 1992]. The actual shape is determined by the properties of the resin and laser. The part building process is assumed to be a stacking up of rectangular shaped blocks. However, the actual process employs different shaped blocks. As a result of the shape of scanned lines, a supposedly vertical edge is not a straight line, but is rather jagged (see Figure 9.9); furthermore, the stair-steps are not well shaped, because their corners are deformed and do not fall on a straight line. This type of error also affects curves (see Figure 9.10, a built feature with a circular section).

Figure 9.8 Scanned line cross section

Figure 9.9* Vertical wall

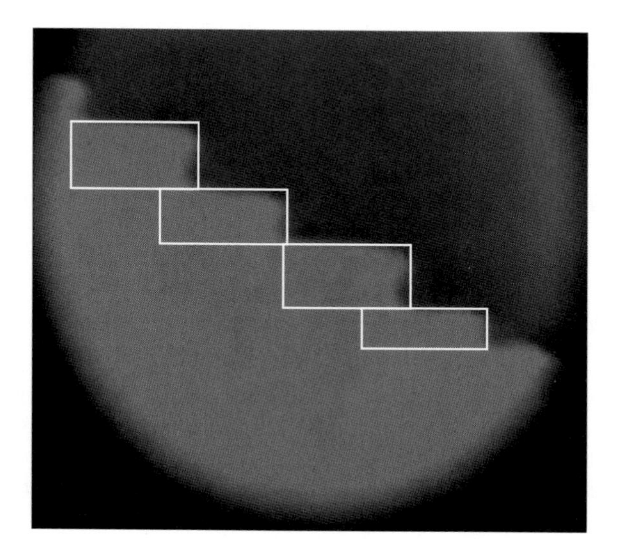

Figure 9.10* Effects on curves

- *Control errors.* Theoretically, the layer thickness should be at the defined value and the border line should be positioned at the specified positions. In fact, the layer thickness is variable and the border position is not precise. Figure 9.10 shows the phenomenon of uneven layer thickness. Figure 9.11 gives measurements taken of a built cylindrical feature with a diameter of 4.47 mm. The specified layer thickness was 0.1 mm. The measured data in column 2 reveals the error in layer thickness control. The chord lengths at different layers were also measured and compared with the theoretical values. The data in column 5 show errors of the chord lengths that reflect the error in border position control.

Layer	Layer thickness measured (mm)	Chord length measured (mm)	Chord length expected (mm)	Errors of chord length (mm)
1	0.084	1.725	1.557	-0.168
2	0.1295	1.992	2.015	0.023
3	0.0755	2.257	2.37	0.113
4	0.123	2.753	2.663	-0.09
5	0.0795	2.961	2.914	-0.047
6	0.1225	3.16	3.131	-0.029
7	0.075	3.378	3.323	-0.055
8	0.1335	3.563	3.492	-0.071
9	0.0725	3.678	3.643	-0.035
10	0.1145	3.8	3.777	-0.023

Figure 9.11 Measured layer thickness and chord length

9.1.2.2 Part Building Errors in the SLS Process

Many materials can be processed by SLS (see Chapter 4). All of them have different properties and characteristics that can affect part accuracy. The main cause of part inaccuracy is the shrinkage during sintering which does not always occur in a uniform manner. The shrinkage of a new layer can be constrained by the existing part substrate or by support powder trapped within enclosed areas. In addition, areas at high temperatures tend to shrink more than those at lower temperatures and part geometries such as thick walls or sections can increase the shrinkage [Beaman et al., 1997]. To compensate for shrinkage, a material shrinkage coefficient is calculated using a test part and a scaling factor is applied in each direction to the STL file. However, scaling a three-dimensional faceted file uniformly is not a simple task and the resulting geometry can be slightly deformed compared to the nominal geometry depending on the scaling technique used [Beaman et al., 1997].

The best way to overcome problems due to shrinkage would be to build a trial part, measure its shrinkage and distortion and rebuild it according to a new design taking into account the dimensions and shape changes that occur during the process. Unfortunately, this approach is unrealistic because it is time consuming and costly. In practice, to compensate for the shrinkage, scaling and offsetting are applied to the part dimensions according to the following relation:

$$\text{new dimension} = a\,(\text{desired dimension}) + b$$

In this way, it is possible to compensate for the shrinkage occurring during the SLS process and for the part growth due to the laser beam melting diameter (Figure 9.12).

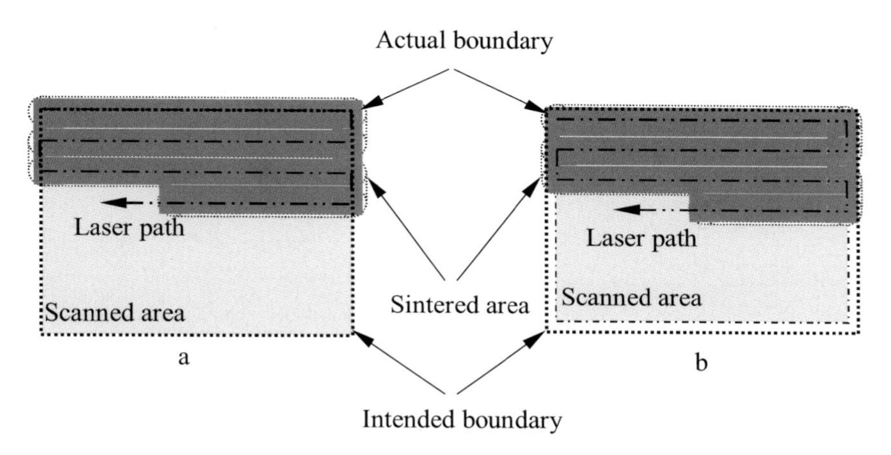

Figure 9.12 Shrinkage and laser beam sintering diameter
(a - without compensation, b - after shrinkage and offset compensation)

The normal procedure to determine the scaling factor a and offset value b consists of building a test part and tabling measurements of it. From these measurements, values a and b are calculated for the X and Y axes (which are normal to the build direction) assuming linear shrinkage for the SLS process.

For example, to calibrate the SLS process for the RapidSteel 2.0 material a test part has to be built (Figure 9.13). This test part has internal and external features in order to determine accurately the scaling factors and the offset values. After building the test part, dimensions in the X and Y directions are measured (Figure 9.14). From these measurements, the scaling factors and the offset values are calculated (Figure 9.15).

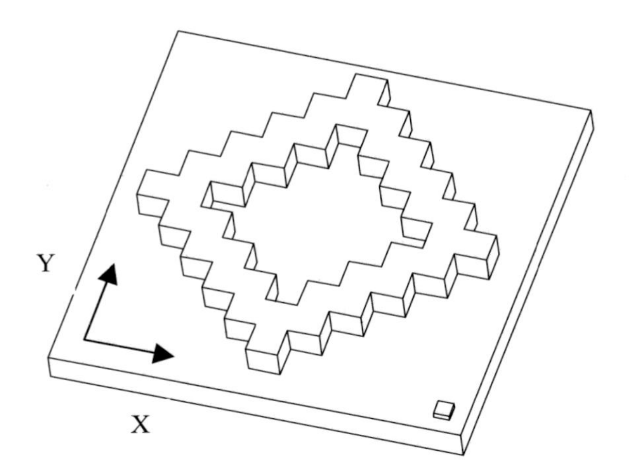

Figure 9.13 RapidSteel 2.0 calibration test part

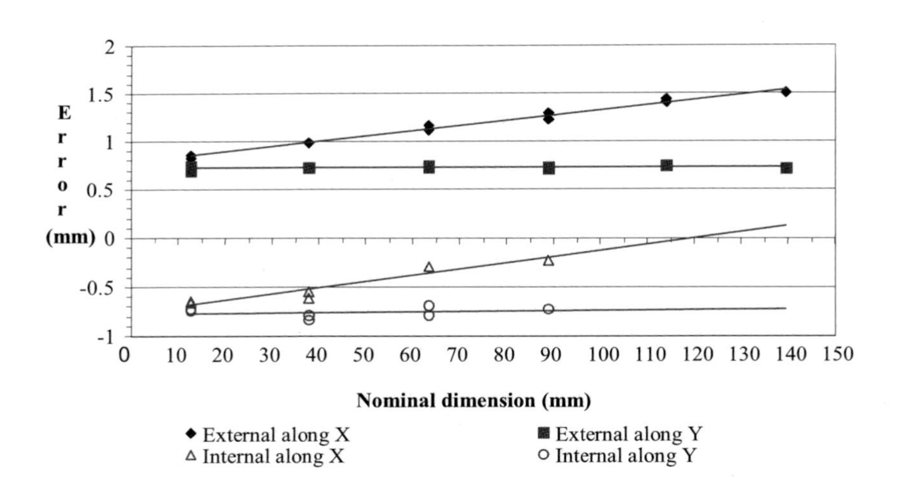

Figure 9.14 Error versus nominal dimension for the test part along the X and Y axes

	External Dimensions		Internal Dimensions	
	X	Y	X	Y
Scaling factor	0.9946	1.0000	0.9937	0.9997
Offset Value	0.3897	0.3608	-0.3786	-0.3861

Figure 9.15 Scaling factors and offset values calculated for the test part

Apart from shrinkage effects that can be different in each direction, the accuracy along X and Y is directly dependent on the scanning system utilised by the machine and its capability to generate the desired laser path. The laser beam positioning is accomplished with mirror galvanometers and errors will be related to the time required to accelerate and decelerate them [Beaman et al., 1997].

The scaling factor for the Z axis (the build direction) is assumed constant and is given by the manufacturer. The accuracy in the Z direction is affected by two problems. The first problem, common to all LMTs, is that some part dimensions in the Z direction might have values different from a multiple of the layer thickness. The second problem is common to the SL and SLS processes. With both processes, the bonding between the layers is achieved by applying a laser power sufficient for the layer being built to bond with the preceding layer. The problem arises with downward-facing layers. As they have no layer underneath them, they are slightly thicker, which generates dimensional errors [Chen and Sullivan, 1995]. New software is now available which allows this problem to be compensated for.

9.1.3 Part Finishing

Some RP applications such as fabrication of exhibition quality models, tooling or master patterns for indirect tool production (see Chapter 8) require additional finishing to improve the surface appearance of the part. To achieve this, the stair-step effect on important surfaces has to be removed. Usually, this is done by sanding and polishing RP models, which leads to changes in feature shapes, dimensions and positions. The model accuracy after finishing operations is influenced mostly by two factors, the varying amount of material that has to be removed and the finishing technique adopted. These two factors determine to what extent the dimensional accuracy of RP models will be reduced during finishing.

- *Varying amount of material.* During the data preparation stage, the RP model shapes are approximated with the corners of the stair-steps. Each RP process reproduces the corners and the stair-steps with a different resolution. Hence, the amount of material that has to be removed to improve the surface finish will vary depending on the RP process employed. Also, the amount of material to be removed on surfaces of the same model can vary due to the selected part build orientation. For example, Figure 9.16a shows a surface profile of a SL model where the stair-steps can be seen. Figure 9.16b depicts another surface of the same model but with a minimal stair-step effect. The former was built facing upwards and the latter downwards. As a result of the part build orientation, varying amounts of cured resin are left for removal around the periphery. In practice, it is very difficult to remove a varying amount of material from the surfaces of the same model without reducing the model accuracy.

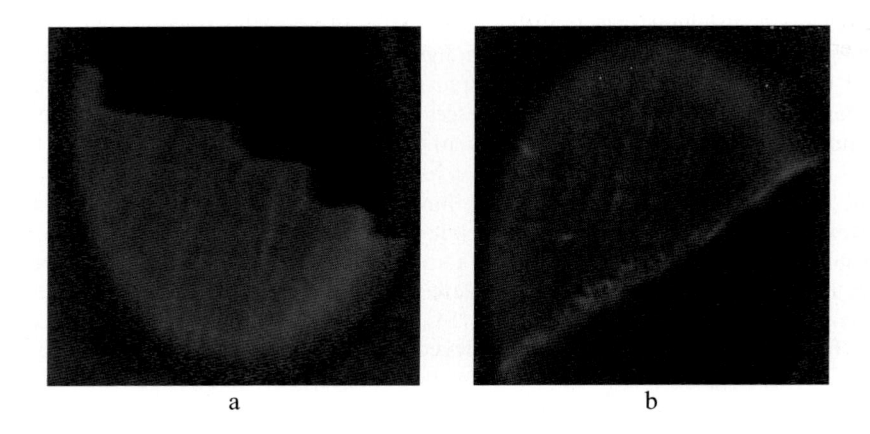

Figure 9.16* Stair-steps around the periphery
(a - facing upwards, b - facing downwards)

- *Finishing technique.* A number of processes can be employed to finish RP models, for example, wet and dry sanding, sand blasting, coating, spraying, infiltration with special solutions, machining, etc. Each technique has specific technological capabilities and can be characterised by the achievable dimensional accuracy and surface roughness. The techniques that assure better dimensional control during the finishing operation will have less impact on model accuracy. For instance, models finished employing milling will have less influence on accuracy than those using manual wet sanding or sand blasting.

9.2 Selection of Part Build Orientation

One of the most important decisions to be made when employing any particular RP technology is the selection of the part build orientation. This decision is a very important factor in minimising build time and costs, and achieving optimal accuracy. When making this decision, designers and RP machine operators should consider a number of different process specific constraints. This may be quite a difficult and time-consuming task.

Each RP process has specific technological capabilities that have to be taken into account before a particular build direction is selected. Choosing the best orientation is a multi-criteria task that involves trade-offs between maximising the surface smoothness and accuracy of important features and minimising the build time and cost. In this section, the orientation constraints of two RP processes, SL and SLS, are discussed. Because SL and SLS have orientation constraints that are common with other RP methods, the information included in this section can be regarded as generic for layer manufacturing technologies.

9.2.1 Orientation Constraints of the SL Process

The SL process has specific constraints that have to be taken into account in identifying possible part-build orientations. These constraints are defined by the technological capability of the process, in particular the achievable build accuracy in the X-Y plane and Z direction, and the necessity for support structures for overhanging areas. To maximise the accuracy of critical areas of a SL model, its build direction has to be chosen to suit particular part features. As a consequence of this, other features may not be optimally orientated, which can further reduce their surface finish or accuracy. For example, if a critical surface is chosen to be horizontal and upward facing, this may cause several holes to be constructed with horizontal axes (axes lying in the X-Y plane), resulting in internal support structures and poor surface finish.

The following feature constraints should be considered in choosing candidate build orientations for the SL process [Pham et al., 1999].

- *User specified critical surfaces*: If these surfaces are planes, they have to be placed such that their normals point in the build direction. In other words, they are horizontal and upward facing. Cylinders, cones and surfaces of revolution are orientated so that their axes are vertical.

- *Coordinate systems*: Since a coordinate system is usually created by the designer and employed whilst modelling, the orientation of the coordinate axes may represent the most logical build direction. It is placed so that the z-axis points in the build direction.

- *Holes:* In order to avoid hard-to-remove supports and stair-stepping inside holes, these are placed orthogonally to horizontal planes.

- *Cuts:* If these carve through the part entirely or have a depth greater than a certain minimum, the planes which they cut through (placement planes) are made horizontal. Otherwise, they are ignored.

- *Shafts:* These are orientated such that their placement planes are horizontal in order to provide the best external surface finish.

- *Protrusions:* If these are created by revolving a section, the axes are positioned so that they are vertical.

- *Shells:* These are orientated so that the concave part of the shell faces upwards in order to minimise internal supports. However, if the part is built on an older SL system employing the deep-dip recoat method, this orientation should be avoided as it would produce a trapped volume [Pham et al., 1997].

- *Axes:* All axes are placed so that they are vertical.

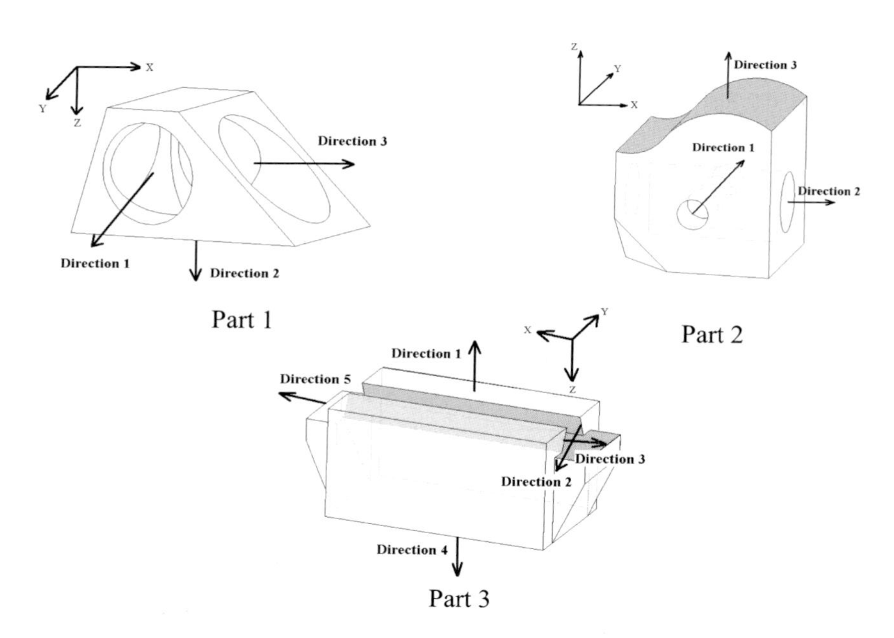

Figure 9.17 Three simple parts used to illustrate the selection of a build direction

Taking into account the feature constraints listed above, candidate orientations can be identified. Each candidate orientation can then be evaluated based on the number of "problem" features, i.e. features not optimally built due to the selected orientation, the amount of support structures required and the build time [Pham et al., 1999]. The principles involved in evaluating competing part orientations can be illustrated on three simple parts shown in Figure 9.17. In Figure 9.18, the evaluation results are presented.

Part Orientations	Part 1	Part 2	Part 3
Build Direction (Z-vector)	0,1,0	0,1,0	0,0,-1
Time	52m	1hr 15m	22hr 36m
Critical Features	2[axis,hole]	2[surface, axis]	1[surface]
Problem Features	2[axis,hole]	2[cut,axis]	3[2 surfaces,cut]
Overhanging Area	31.71	114.43	6000
Support Volume	8.57	37.22	2250
Build Direction (Z-vector)	1,0,0	1,0,0	0,-0.91,0.42
Time	1hr 12m	1hr 15m	24hr 42m
Critical Features	2[hole,axis]	2[axis,hole]	1[surface]
Problem Features	2[hole,axis]	1[axis]	3[2 surfaces,cut]
Overhanging Area	25.47	120.26	6802.84
Support Volume	6.62	32.54	6557.12
Build Direction (Z-vector)	0,0,1	0,0,1	0,0.91,0.42
Time	52m	1hr 19m	24hr 42m
Critical Features	1[Coord. Sys.]	1[Coord. Sys.]	1[surface]
Problem Features	4[2 holes, 2 axes]	3[2 holes, axis]	3[2 surfaces,cut]
Overhanging Area	68.71	126.11	6802.84
Support Volume	28.92	38.8	6557.12
Build Direction (Z-vector)			0,0,1
Time			22hr 48m
Critical Features			1[Coord. Sys.]
Problem Features			4[3 surfaces,cut]
Overhanging Area			8051.75
Support Volume			4783.98
Build Direction (Z-vector)			1,0,0
Time			27hr 57m
Critical Features			1[cut]
Problem Features			3[surfaces]
Overhanging Area			2929.68
Support Volume			2315.08

Figure 9.18 Evaluation results for the competing build orientations

9.2.2 Orientation Constraints of the SLS Process

The orientation constraints of the SLS are dictated mostly by two factors, the process accuracy in the X-Y plane and Z direction, and the material used, since the shrinkage and anisotropic properties of the sintered powders are different. To utilise the technological capabilities of the SLS process fully, designers and machine operators should understand its specific part orientation rules.

Commercially available SLS materials fall into the following five categories [Pham et al., 1997]: **Polycarbonates, Polyamides, Elastomers, Metals and Sands**. Different materials impose different orientation constraints on design features (Figure 9.19). Therefore, the orientation constraints are described below along with the materials classes to which they apply [DTM, 1995; DTM, 1996a; DTM, 1996b; DTM, 1998a; DTM, 1998b; DTM, 1998c].

- To maximise strength, snaps and pegs are built in the X-Y plane so that they will have improved flexibility (Figure 9.19). This is because the finished prototypes have anisotropic properties due to the scanning direction and the layering effect. This rule should be applied when building with **polycarbonates, elastomers** and **polyamides**.

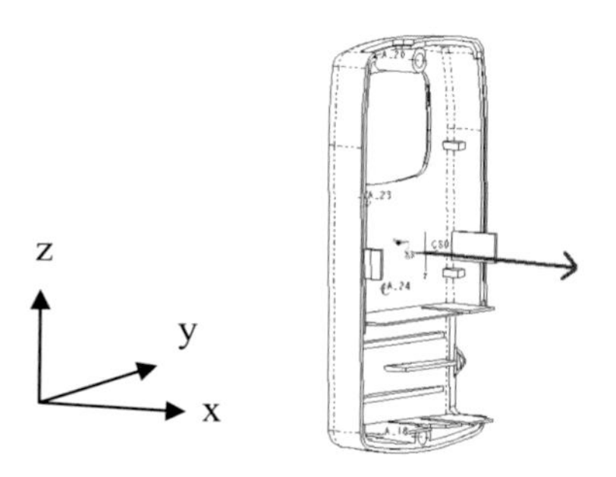

Figure 9.19 To maximise strength, fine features should be built in the X-Y plane

- For the best feature definition, critical features should be built on upwards facing model surfaces (Figure 9.20). This should be done when building with **sand**, **elastomers** and **polyamides**.

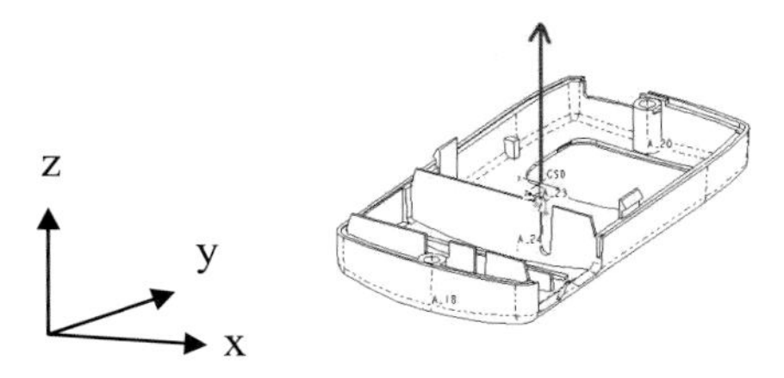

Figure 9.20 Fine features should be built facing upward to achieve better feature definition

- With **polycarbonates**, it is important that the rate of change of cross-sectional area is not too rapid to ensure consistent shrinkage. Thus parts are normally angled by 10 - 15 degrees around the X axis. After an orientation has been chosen, the user is advised to slice the part to check that this rate of change is acceptable.

- When creating parts with **sand**, it is necessary to build the prototype in the same orientation as that in which it will be cured to reduce warping. Therefore the part must be stable in the build direction.

- Thin walls should be built perpendicular to the roller's direction of travel to avoid their distortion when building with **polycarbonates** and **elastomers** (Figure 9.21).

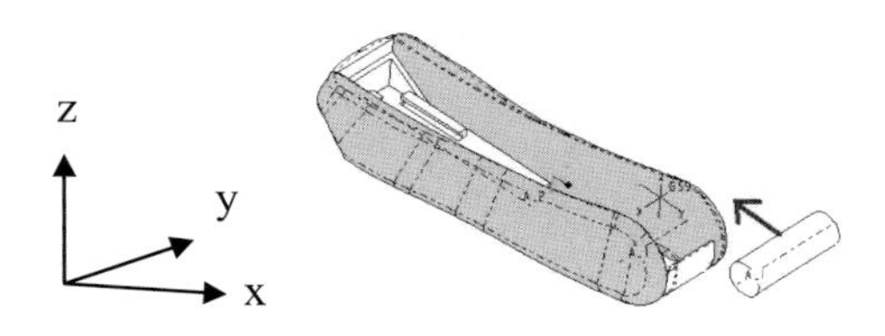

Figure 9.21 Thin walls should be constructed perpendicular to the roller

- Shells should be built so that the concave side points upwards to reduce trapped heat and minimise growth (Figure 9.22). This rule should be applied when building with **polycarbonates**.

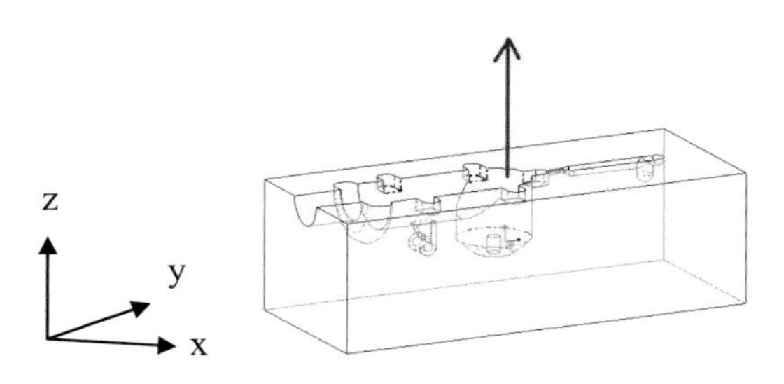

Figure 9.22 The concave side of shells should point upward to reduce trapped heat and minimise growth

- Overhangs should be minimised. This rule should be employed when building with **polycarbonates**, **sand** and **elastomers**.

- When making **metal** tooling inserts, the build direction should be chosen such that the parting line of the mould is built facing upwards.

The constraints listed above can be used to identify possible build directions. As is the case with the SL process, the final selection of the build direction is carried out by evaluating each candidate orientation based on the number of "problem" features, i.e. features not optimally built due to the selected orientation and build time [Pham et al., 2000].

9.3 Summary

Proper utilisation of RP processes requires a thorough understanding of their capabilities and limitations. This chapter provides an overview of the main factors influencing RP part accuracy together with process-specific constraints that should be considered in selecting part build orientations. Choosing the optimal process parameters and build orientations involves a trade-off between maximising part accuracy and surface finish and minimising build height and cost. The issues discussed in this chapter give an insight into the considerations associated with the optimal use of RP processes.

References

Beaman JJ, Barlow JW, Bourell DL, Crawford RH, Marcus HL and McAlea KP, (1997) Solid Freeform Fabrication: A New Direction in Manufacturing, **Kluwer Academic Publishers**, Dordrecht, The Netherlands.

Chen CC and Sullivan PA (1995) Solving the Mystery - The Problem of Z-Height Inaccuracy of the Stereolithography Parts, **Proceedings of 6th International Conference on Rapid Prototyping**, June, Dayton, Ohio, pp 153-170.

DTM Corporation (1995) Nylon: Guide to Materials, **DTM Corporation**, 1611 Headway Circle, Building 2, Austin, Texas 78754.

DTM Corporation (1996a) TrueForm: Guide to Materials, **DTM Corporation**, 1611 Headway Circle, Building 2, Austin, Texas 78754 .

DTM Corporation (1996b) Polycarbonates: Guide to Materials, **DTM Corporation**, 1611 Headway Circle, Building 2, Austin, Texas 78754.

DTM Corporation (1998a) Sandform: Guide to Materials, **DTM Corporation**, 1611 Headway Circle, Building 2, Austin, Texas 78754.

DTM Corporation (1998b) Somos 201: Guide to Materials, **DTM Corporation**, 1611 Headway Circle, Building 2, Austin, Texas 78754.

DTM Corporation (1998c) RapidSteel: Guide to Materials, **DTM Corporation**, 1611 Headway Circle, Building 2, Austin, Texas 78754.

Jacobs, PF (1992) Rapid Prototyping and Manufacturing, Fundamentals of stereolithography, **Society of Manufacturing Engineers**, Dearborn, MI.

Pham DT, Dimov SS and Gault RS (1997) A Guide to Prototyping, **Prototyping Technology International '97**, UK and International Press, Surrey, UK, pp 15-19.

Pham DT, Dimov SS and Gault RS (1999) Part Orientation in Stereolithography, **Int J Adv Manuf Technol**, Vol. 15, pp 674-682.

Pham DT, Dimov SS and Gault RS (2000) Part Orientation in Selective Laser Sintering, Internal Report, **Manufacturing Engineering Centre**, University of Wales Cardiff, Cardiff, UK.

Pham DT and Ji C (2000) Design for Stereolithograthy, **Proc. IMechE, Part C**, Vol. 214, pp 635-640.

Rause FL, Ulbrich A, Ciesla M, Klocke F and Wirtz H (1997) Improving Rapid Prototyping Speeds by Adaptive Slicing, **Proceedings of the 6th European Conference on Rapid Prototyping and Manufacturing**, July, Nottingham, pp 31-36.

Reeves PE, Dickens PM, Davey N and Cobb RC (1997) Surface Roughness of Stereolithography Models Using an Alternative Build Strategy, **Proceedings of the 6th European Conference on Rapid Prototyping and Manufacturing**, July, Nottingham, pp 85-94.

Williams RE, Komaragiri SN, Melton VL and Bishu RR (1996) Investigation of the Effect of Various Build Methods on the Performance of Rapid Prototyping (Stereolithography), **Journal of Materials Processing Technology**, Vol. 61, 1-2, pp 173-178.

Author Index

Subject Index